D1367557

MILITARY OUTLOOK

DR. ANIS I. MILAD

authorHOUSE®

AuthorHouse™
1663 Liberty Drive
Bloomington, IN 47403
www.authorhouse.com
Phone: 1 (800) 839-8640

The cover image is provided by the international artist, Merna Amged.

Published by AuthorHouse 05/15/2019

ISBN: 978-1-7283-1215-6 (sc)
ISBN: 978-1-7283-1214-9 (e)

Print information available on the last page.

This book is printed on acid-free paper.

CONTENTS

PART 1

PART 2

PART 1

ANTI-SUBMARINE WARFARE

COPYRIGHT 2015

ANTI-SUBMARINE WARFARE

The Problem:

Delivering the unmanned vessel to the enemy activity zone is slow and not practical. The quiet submarines are not detected in the deep water.

The Solution:

The solution is to launch geographical device to orbit the earth and locate all the submarines in the oceans. This technology is already used for geographical purposes. Taking inventory for all submarines which are in the oceans and the seas is important for protecting this country. The NAVY will launch the satellite archaeology "Some 400 miles up in space, satellites collect images that are used to identify buried landscapes with astonishing precision." (Bloch, 2013). "University of Alabama at Birmingham archaeologist Sarah Parcak, is a pioneer in using satellite imagery in Egypt." (Bloch). The satellite archaeology is not limited to buried landscapes.

The Idea:

1) The geological device in the orbit will allow the Navy to locate all the submarines which are in the oceans and the seas either quiet submarines or not quiet submarines.
2) The location of each submarine continually will be known to the Navy.
3) The quiet submarine will be distinguished and detected because we already know the location of the noisy submarines.

4) Two options are remained. The enemy, quiet or not quiet, submarine will be destroyed by the Air Force or the NAVY warfighters or by the unmanned vessels

5) The stealth unmanned vessels are delivered by the Improved Flying Vehicle or the aircraft carrier if it is approachable.

6) The aircraft carrier will carry aboard the stealth unmanned vessel.

7) The Improved Flying Vehicle (IFV) is equipped to fly in the atmosphere.

8) The IFV will be launched from any military base around the globe regardless of the distance between the military base and the enemy's submarine.

9) The IFV will be operated by a pilot(s) who will drive the IFV away from the war zone or the enemy's submarine faster than the speed of sound after releasing of the stealth unmanned vessels is completed.

10) The unmanned vessel's remote navigator will be directed to follow, detect, or destroy the enemy quiet submarine.

11) The IFV or the aircraft carrier could have one or more unmanned vessels.

Reference

Bloch, Hannah (2013) The New Age of Exploration. Satellite Archaeology. Retrieved November 29, 2014 from http://ngm. nationalgeographic.com/2013/02/125-explore/satellite-archaeology

ADJUSTABLE ROCKET ENGINE

COPYRIGHT 2015

ADJUSTABLE ROCKET ENGINE

The Problem:

There are four concepts must be addressed:

1. The first concept is that the speed/maneuver of the rocket engine is not controllable from the start when the rocket engine is ignited and when the rocket engine replaces the jet engine in a warplane.
2. The second concept is associated with the fighter jet, that is, the limited speed of the fighter jet and the huge distance between the American military bases and the location of the enemy.
3. The third concept is that the response of the fighter jet to the satellite signals (pictures and location) is slow when the satellite locates a particular location of the military equipment/ convoy of the enemy because the slow response is due to the huge distance between the USA military bases and the location of the enemy.
4. The fourth concept is that the fighter jet in general confirms limited capability of the pilot as a human being and the maximum/limited speed of the fighter jet.

The Solution:

1. The main concern is to narrow the response time between the signals of the satellite and destroying the enemy. The speed

of Blackbird is Mach 3 but the speed of the rocket is Mach 10 (China is experimenting with Mach 12).

2. When producing the rocket in the same way the Blackbird is produced to be able to carry several bombs/guided missiles/ clusters, intelligent data, stealth, and satellite networking connections the delivery of the bombs for several enemy locations will be faster and instant.

3. The suggested warplane rocket will be unmanned so the physical pilot factor will be eliminated.

4. As an alternative, the warplane rocket will be operated remotely via the satellite.

5. As another alternative, the warplane rocket will be operated by a pilot who will be wearing a space suit and inside a rocket design cockpit.

6. The rocket engine which is suggested by this proposal for this warplane rocket must be equipped with adjustable pumps to control the speed of the rocket and its maneuvers (please see the drawing).

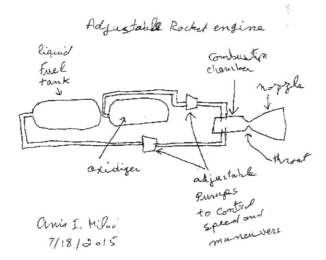

Adjustable Rocket engine

AIRBORNE HYBRID VEHICLE

COPYRIGHT 2013

AIRBORNE HYBRID VEHICLE

The Problem:

"Satellites today are launched via booster rocket from a limited number of ground facilities, which can involve a month or longer of preparation for a small payload and significant cost for each mission. Launch costs are driven in part today by fixed site infrastructure, integration, checkout and flight rules. Fixed launch sites can be rendered idle by something as innocuous as rain, and they also limit the direction and timing of orbits satellites can achieve." (TTO Programs)

The Solution:

The solution is to combine a rocket and aircraft in one flying vehicle. This improved flying vehicle is equipped to fly in the atmosphere and to orbit the earth in space. The reason this flying vehicle should be hybrid is to be able to orbit the earth and to shorten the time/costs to deliver the Satellite.

The Idea:

1) To create an improved flying vehicle (IFV) to carry the satellite to the orbit.
2) The rocket of the IFV will have a new nozzle design.
3) The IFV will be launched from any military base around the globe regardless of the bad weather here or in other part of the world because the IFV will be launched to the earth orbit in a few minutes in a clear sky or not as an ordinary jet aircraft.

4) The IFV will be operated by a pilot(s) who will drive the IFV using jet engines then will ascend to orbit via the rocket.

5) The two jet engines will be turned off and closed from both side while the IFV is in the orbit.

6) The rocket will be fueled for the exact distance from the moment the two jet engines are turned off and the moment the rocket is ignited.

7) The IFV will start by flying upward as an ordinary jet airplane.

8) The rocket will be ignited to overcome the gravity in its way to the orbit.

9) The IFV will be reused indefinitely as long as no maintenance is required.

10) The rocket will be refueled with solid fuel or other suitable fuel.

11) Two options are remained for the IFV which are to release a new satellite or to upload an old satellite for repair or because it is outdated.

United States Patent [19]

Milad

[11] **Patent Number:** 4,732,106

[45] **Date of Patent:** Mar. 22, 1988

[54] **STEERING CONTROL FOR SUBMARINES AND THE LIKE**

[76] Inventor: Anis L. Milad, 2938 Yorkway, Baltimore, Md. 21222

[21] Appl. No.: 887,921

[22] Filed: Jul. 22, 1986

[51] Int. Cl.⁴ B63G 8/16; B63H 5/12
[52] U.S. Cl. 114/338; 74/89.15;
74/665 B; 219/121 LU; 239/265.35; 239/587;
244/51; 244/52; 440/58; 440/63; 901/25
[58] Field of Search 114/337, 338;
440/58–60, 63; 244/51, 52, 66, 76 J, 169, 74;
239/265.35, 587; 74/665 B, 665 A, 89.15;
901/25; 219/121 LU, 121 LV, 121 LW, 121
LX

[56] **References Cited**

U.S. PATENT DOCUMENTS

1,409,850	3/1922	Haney	244/51 X
2,131,155	9/1938	Waller	244/51
3,809,318	5/1974	Yamamoto	239/587 X
4,281,795	8/1981	Schweikl	239/265.35
4,497,319	2/1985	Sebate et al.	219/121 LU
4,501,322	2/1985	Causer et al.	901/25 X
4,542,278	9/1985	Taylor	219/121 LV
4,578,554	3/1986	Coulter	219/121 LV
4,579,299	4/1986	Lavery et al.	239/265.35 X

Primary Examiner—Sherman D. Basinger
Attorney, Agent, or Firm—Leonard Bloom

[57] **ABSTRACT**

A steering apparatus for steering propulsion devices such as a propeller shaft and rocket exhaust and for aiming devices such as a fire hoses and lasers. The steering apparatus provides a rotational component and a tilting component to the orientation of the steered device.

13 Claims, 16 Drawing Figures

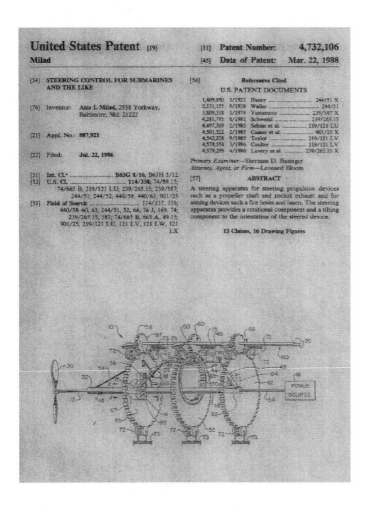

US005119753A

United States Patent [19]

Milad

[11] Patent Number: 5,119,753

[45] Date of Patent: Jun. 9, 1992

4,732,106	3/1988	Milad		440/63 X
4,907,937	3/1990	Milenkovic		901/29 X

[54] ARTICULATABLE MECHANISM

[76] Inventor: Anis I. Milad, 2938 Yorkway, Baltimore, Md. 21222

[21] Appl. No.: 569,407

[22] Filed: Aug. 20, 1990

[51] Int. Cl.^{5} B63G 8/16

[52] U.S. Cl. .... 114/338; 74/665 B; 901/29; 440/63

[58] Field of Search .... 901/18, 25, 29; 74/661, 74/665 B, 665 L, 665 G, 665 N; 244/51, 52, 66, 76 J; 475/1, 2; 239/265.35, 587; 114/338, 337; 440/58–63, 53

[56] **References Cited**

U.S. PATENT DOCUMENTS

1,409,850	3/1922	Haney		244/51 X
4,501,522	2/1985	Causer et al.		901/25 X

Primary Examiner—Edwin L. Swinehart
Attorney, Agent, or Firm—Leonard Bloom

[57] **ABSTRACT**

This invention is an articulatable mechanism applicable to a wide variety of situations. The device provides a rotational component as well as a pivotal component providing practical propulsion in polar coordinates instead of customary linear transmission. This device could be used for steering propulsion devices by controlling the orientation and direction of a propeller shaft or rocket exhaust. Another application of this device might be orienting and directing devices such as fire hoses and lasers.

12 Claims, 9 Drawing Sheets

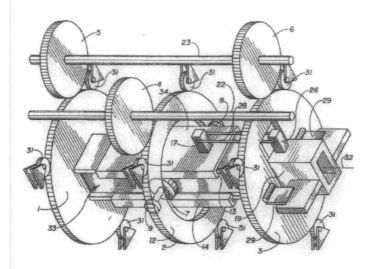

ALTERNATIVE METHODS TO EXTINGUISH THE FOREST FIRE

COPYRIGHT 2013

ALTERNATIVE METHODS TO EXTINGUISH THE FOREST FIRE

Updated

The Problem:

The fire in the forest spreads quickly without warning. The fire is caused in the nature for two reasons. The first reason is the lightning, and the second reason is the act of terrorists.

The Solution:

The first solution is to install fire alarms that are sensitive to heat, smoke, and sound. The second solution is to install video cameras.

The Idea:

1) The fire alarms will be installed randomly throughout the forest.
2) The fire alarms will be hidden and very difficult to detect and invisible to the eyes.
3) The fire alarms will be connected as a network.
4) The location of each fire alarm in the fire alarms network will be known on the forest map.
5) The fire alarm network will be connected to the fire stations computers in the state.

6) The location will be determined and the anti-fire airplanes will respond to the fire alarms.

7) The fire alarms will be designed to absorb/distinguish the intense heat and smoke that are caused by the forest fire.

8) The video cameras will be activated by the fire/smoke/sound and send instant videos to the fire stations computers.

9) The fire alarms and the video cameras will be operated by super batteries.

10) The fire alarms and the video cameras will be installed by technicians who will be using helicopters to reach the out of reach locations in the forest.

11) The fire alarms will be designed to absorb the plunge force on the ground when the technician chooses to throw them from the helicopter/airplane. Instead the technician is lowered down to the forest floor to install the fire alarms s/he could throw them from the helicopter/airplane to the designated locations on the forest map.

12) The fire alarms and the video cameras will be resistant to the water and the chemical that is used to extinguish the fire.

13) The fire alarms and video cameras will be replaced when it is necessarily.

ALTERNATIVE WEAPONS

COPYRIGHT 2015

ALTERNATIVE WEAPONS

The Problem:

1) Hitting the target and escaping the war zone will not be accomplished easily because there is a strong possibility the enemy's warplanes will follow our warplane and they will shoot it down.
2) Firing ground-to-air missiles against our warplanes will be precise and deadly.
3) Operating unmanned or manned missiles through the satellite or through manual defense systems must need different strategies to lessen the effect of the counter attacks.

The Solution:

The Strategies are:

1) The first strategy is to equip our warplanes with laser beam which will be aimed toward the enemy's warplane that follows our warplane after hitting the target in the enemy's territory.
2) The purpose of the laser beam is to be used as a destruction force and/or to blind the pilot of the enemy's warplane.
3) The second strategy is that if firing ground-to-air missiles against our warplanes is done through the satellite and automatically, our warplane must aim high frequency electromagnetic signals against the enemy's defense missiles' location to create electrical interference to the enemy's defense electronic equipment.

4) As another alternative, drones could be specialized to produce high frequency electromagnetic signals by flying in lower altitude than our warplane which is flying in higher altitude to attack the ground enemy location in the war zone.

5) Also the high frequency electromagnetic signals could be used against the enemy's guided missiles to create electrical interference to the enemy's guided missiles electronic system.

6) In the sea and ocean, creating small quiet submarines that are specialized to produce high frequency electromagnetic signals could be the solution to cripple the enemy's quiet submarines when creating electrical interference in the enemy's quiet submarines.

7) Exposing the quiet submarines should be done when launching geographical device to orbit the earth and locate all the submarines in the oceans. This technology is already used for geographical purposes. Taking inventory for all submarines which are in the oceans and the seas is important for protecting this country. The NAVY will launch the satellite archaeology "Some 400 miles up in space, satellites collect images that are used to identify buried landscapes with astonishing precision." (Bloch, 2013). "University of Alabama at Birmingham archaeologist Sarah Parcak, is a pioneer in using satellite imagery in Egypt." (Bloch). The satellite archaeology is not limited to buried landscapes.

8) Creating drones and small submarines that are equipped/ specialized to produce high frequency electromagnetic signals could be the solution to cripple the enemy's quiet submarines, guided missiles, air-to-air missiles and ground-to-air missiles when creating electrical interference in the enemy's electronic offense/defense systems.

Reference

Bloch, Hannah (2013) The New Age of Exploration. Satellite Archaeology. Retrieved November 29, 2014 from http://ngm. nationalgeographic.com/2013/02/125-explore/satellite-archaeology

ALTERNATIVES FOR CONVENTIONAL ENCOUNTER

COPYRIGHT 2013

ALTERNATIVES FOR CONVENTIONAL ENCOUNTER

Description: Helicopter Tank is a combination of a Helicopter and Tank. It consists of helicopter, tank gears, and a removable Howitzer cannon. The Helicopter Tank is designed to fly low, to avoid the radar, and to be driven on the ground in a conventional encounter with the enemy. During the peace time the Helicopter Tank will be used to search for oil and natural gas in the desert and elsewhere. The tank gears are made of light hard aluminum or other strong light weight metal. The transmissions of the horizontal and the vertical propellers will be separated from the transmission of the tank gears.

The Idea: During the conventional wars it is necessarily to trick the enemy. The army subunit needs to confuse and to be able to change the position and the location if the fight was during the night. To stay in the same location after a night encounter will expose the army subunit to the enemy air attack in the dawn of the following day especially if the army subunit is a subunit of a howitzers army unit. The loud firing of the howitzers will be another reason for relocating the army subunit after fighting.

The Tactic: The army subunit could substitute its all howitzers with only one howitzer which will be attached to the Helicopter Tank which will be relocated in a temporarily location. In this temporarily location the Helicopter Tank will move quite a few yards each time the howitzer fires. The Helicopter Tank that is equipped by the howitzer will be on the move which gives the impression there are several howitzers.

When the ammunitions are exhausted the Helicopter Tank will, in the night, be driven on the ground to a safe distance and take off and fly back to the subunit's original location. The soldiers of this subunit will leave fake howitzers which look from a distance as real howitzers expecting the enemy air attack in the dawn.

Flexibility: The Captain of the subunit could decide that the howitzer that is attached to the Helicopter Tank be detached and used quietly on the ground to fire the ammunitions or to keep the howitzer attached to the Helicopter Tank while firing it. Shock absorbers will be used for both alternatives. The Helicopter Tank is used instead of a truck to deliver/return the howitzer for the incredible maneuver and speed of the Helicopter Tank.

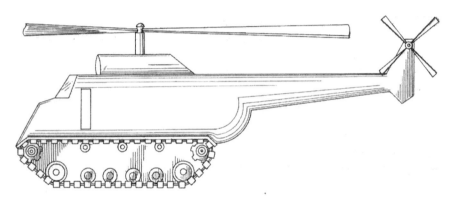

United States Patent

[19] Milad

[11] Patent Number: Des. 292,193 [45] Date of Patent: ** Oct. 6, 1987

[54] COMBINED HELICOPTER AND TANK

[76] Inventor: Anis I. Milad, 29-38 Yorkway, Baltimore, Md. 21222

[**] Term: 14 Years

[21] Appl. No.: 690,587

[22] Filed: Jan. 11, 1985

[52] U.S. Cl. D12/2; D21190; D211I31

[58] Field of Search.......D211131, 85; D12/214, D12/12; 446/230,37,237,433,465; 244/1 R, 2, 17.11, 17.17

1,788,440 l/t931 Prell.
2,019,482 11/1935 Barnes.
2,424,769 7/1947 Page.
2,560,747 7/1951 Sievers.
3,761,040 9/1973 Cummins.

Primary Examiner-Charles A. Rademaker
Attorney, Agent, or Firm-Leonard Bloom

[57]   CLAIM

The ornamental design for a combined helicopter and tank, as shown and described.

[56]   References Cited
U.S. PATENT DOCUMENTS
D. 135,849 6/1943 Gibson...................D21/131
D.138,329 7/1944 Wasson...................D21/85
D.169,229 3/1953 Tober.....................D21/131
D. 284,594 7/1986 Fisher....................D21/85
D.287,378 12/1986 Ohno...................D21/131

DESCRIPTION

FIG. 1 is a side elevational view of a combined helicopter and tank showing my new design, the side opposite being substantially a mirror image;

FIG. 2 is a top plan view thereof;

FIG. 3 is a bottom plan view thereof;

FIG. 4 is a front elevational view thereof; and,

FIG. 5 is a rear elevational view thereof. Portions of the blade members have been shown In fragment im FIGS. 2 and 3 for ease of illustration.

AMPHIBIOUS FIGHTING VEHICLE

COPYRIGHT 2015

"The Defense Advanced Research Projects Agency (DARPA) is calling on innovators with expertise in designing and engineering drivetrain and mobility systems to collaboratively design elements of a new amphibious infantry vehicle, the Fast, Adaptable, Next-Generation Ground Vehicle (FANG)............ Each of the three planned challenges will focus on increasingly complex vehicle subsytems and eventually on the design of a full, heavy amphibious infantry fighting vehicle that conforms to the requirements of the Marine Corps' Amphibious Combat Vehicle (ACV). In the course of the design challenges, participants will test DARPA's META design tools and its VehicleFORGE collaboration environment, with the ultimate goal of demonstrating that the development timetable for a complex defense system can be compressed by a factor of five." (TTO Programs).

The Solution:

To create a Helicoptertank- Submarine.

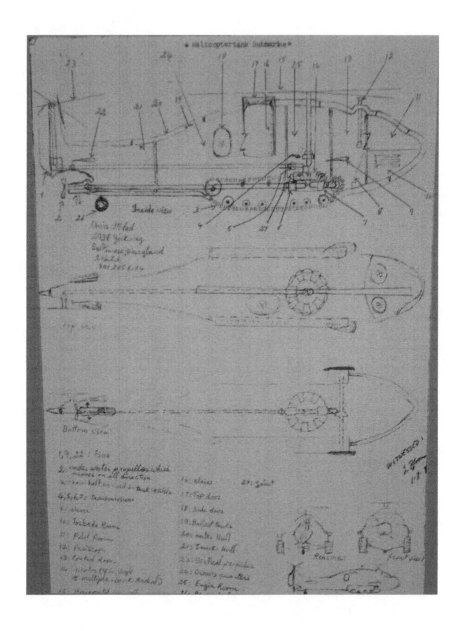

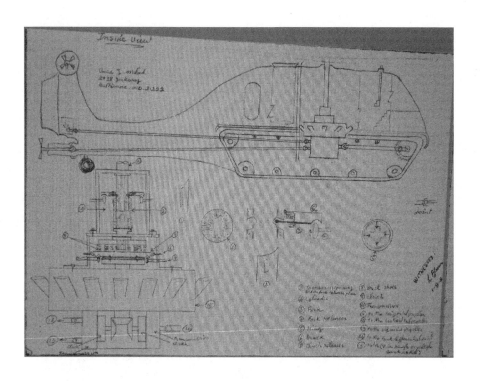

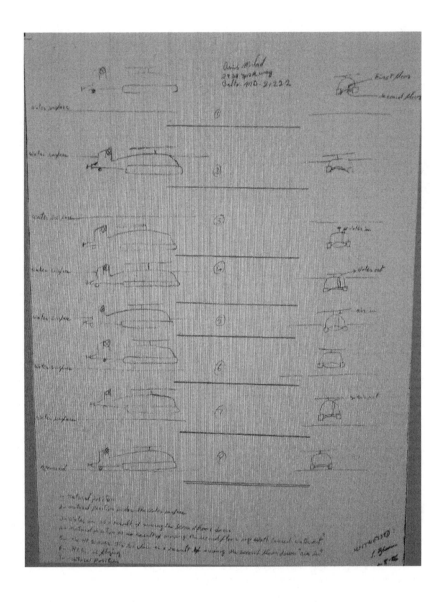

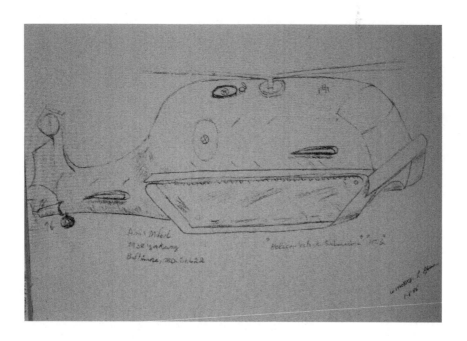

ANTI-SUBMARINE WARFARE

COPYRIGHT 2015

ANTI-SUBMARINE WARFARE

The Problem:

Delivering the unmanned vessel to the enemy activity zone is slow and not practical. The quiet submarines are not detected in the deep water.

The Solution:

The solution is to launch geographical device to orbit the earth and locate all the submarines in the oceans. This technology is already used for geographical purposes. Taking inventory for all submarines which are in the oceans and the seas is important for protecting this country. The NAVY will launch the satellite archaeology "Some 400 miles up in space, satellites collect images that are used to identify buried landscapes with astonishing precision." (Bloch, 2013). "University of Alabama at Birmingham archaeologist Sarah Parcak, is a pioneer in using satellite imagery in Egypt." (Bloch). The satellite archaeology is not limited to buried landscapes.

The Idea:

1) The geological device in the orbit will allow the Navy to locate all the submarines which are in the oceans and the seas either quiet submarines or not quiet submarines.
2) The location of each submarine continually will be known to the Navy.
3) The quiet submarine will be distinguished and detected because we already know the location of the noisy submarines.

4) Two options are remained. The enemy, quiet or not quiet, submarine will be destroyed by the Air Force or the NAVY warfighters or by the unmanned vessels

5) The stealth unmanned vessels are delivered by the Improved Flying Vehicle or the aircraft carrier if it is approachable.

6) The aircraft carrier will carry aboard the stealth unmanned vessel.

7) The Improved Flying Vehicle (IFV) is equipped to fly in the atmosphere.

8) The IFV will be launched from any military base around the globe regardless of the distance between the military base and the enemy's submarine.

9) The IFV will be operated by a pilot(s) who will drive the IFV away from the war zone or the enemy's submarine faster than the speed of sound after releasing of the stealth unmanned vessels is completed.

10) The unmanned vessel's remote navigator will be directed to follow, detect, or destroy the enemy quiet submarine.

11) The IFV or the aircraft carrier could have one or more unmanned vessels.

Reference

Bloch, Hannah (2013) The New Age of Exploration. Satellite Archaeology. Retrieved November 29, 2014 from http://ngm.nationalgeographic.com/2013/02/125-explore/satellite-archaeology

ANTI-TORPEDO SUBMARINES

COPYRIGHT 2015

ANTI-TORPEDO SUBMARINES

The Problem:

A torpedo that is aimed toward an aircraft carrier or another Navy ship must be detected and destroyed before the impact.

The Idea:

The aircraft carrier, for example, is extremely important during the wars and it is protected by missiles against the enemy warplanes but the aircraft carrier is vulnerable when it comes to the torpedoes.

The Solution:

1. Every aircraft carrier must be protected by specialized anti-torpedo submarines. The anti-torpedo submarines will have the capability to dock on both sides of the aircraft carrier.
2. If the anti-torpedo submarines were not docked, the mission is to orbit the aircraft carrier under water to locate incoming enemy torpedoes from a distance away from the aircraft carrier.
3. Both anti-torpedo submarines will continue guarding the aircraft carrier 24/7.
4. The anti-torpedo submarines will be equipped with guided torpedoes which are coded to only destroy the enemy un-coded torpedoes and ships.
5. The anti-torpedo submarines will be carrying stealth torpedoes.

6. The anti-torpedo submarines must be quiet submarines and undetected.
7. The anti-torpedo submarines must be able to sense/detect the enemy torpedoes from all directions including horizontal and vertical incoming enemy torpedoes.
8. During the dock of the anti-torpedo submarines the mission is to replace the used torpedoes with new anti-torpedo torpedoes.

BLACKBOX THE
ALTERNATIVE SOLUTION

COPYRIGHT 2016

BLACKBOX THE ALTERNATIVE SOLUTION

The Problem:

When the airplane crashes in the ocean as a result of an explosion or mechanical reasons, the depth of the ocean and the huge distance of the airplane's pieces from each others will prevent the rescued team for bringing the Blackbox to the surface. Secondly, the battery of the Blackbox is with limited duration.

The Idea:

"The U.S. Navy operates two extremely low frequency radio (ELF) transmitters to communicate with its deep diving submarines.

The Navy's ELF communications system is the only operational communications system that can penetrate seawater to great depths and is virtually jam proof from both natural and man-made interference.

It is a critical part of America's national security in that it allows the submarine fleet to remain at depth and speed and maintain its stealth while remaining in communication with the national command authority.

The Navy's ELF system operates at about 76 Hz, approximately two orders of magnitude lower than VLF. The result is that ELF waves penetrate seawater to depths of hundreds of feet, permitting communications with submarines while maintaining stealth." (Navy Fact File)

Reference: https://fas.org/nuke/guide/usa/c3i/fs_clam_lake_elf2003.pdf

The Solution:

1) Using the extremely low frequency radio (ELF) to communicate with Blackbox.
2) The Blackbox will continue broadcasting the recorded conversation between the pilot and the co-pilot repeatedly when the Blackbox is disconnected from the airplane electrical circles and electrical network..
3) The Naval Computer and Telecommunications Area Master Station – Atlantic will continue to receive the recorded conversation between the pilot and the co-pilot repeatedly.
4) The Blackbox will be equipped with battery and a broadcasting equipment.
5) This solution will eliminate the search for the Blackbox in the ocean.

BREAKING THE BARRIER BETWEEN HUMANS AND COMPUTERS

COPYRIGHT 2015

BREAKING THE BARRIER BETWEEN HUMANS AND COMPUTERS

FLOW CHART FOR CAR X

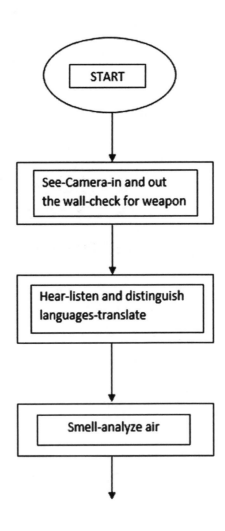

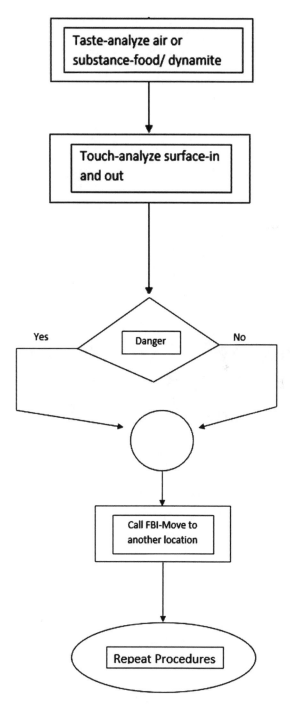

BREAKING THE BARRIER BETWEEN HUMANS AND COMPUTERS

BREAKING THE BARRIER BETWEEN HUMANS AND COMPUTERS

The Problem:

"I2O explores game-changing technologies in the fields of information science and technology to anticipate and create rapid shifts in the complex national security landscape. Conflict can occur in traditional domains such as land, sea, air, and space, and in emerging domains such as cyber and other types of irregular warfare. I2O's research portfolio is focused on anticipating new modes of warfare in these emerging areas and developing the concepts and tools necessary to provide a decisive information advantage for the U.S. and its allies." (Innovation Office (I2O) DARPA-BAA-14-39).

The Outline:

To use the human 6 senses as an outline to create the technology that utilize an ordinary car as shown in the Flow Chart which is provided with this research.

The Solution:

1) This solution is to use the technology on the ground for the national security.
2) To utilize an ordinary car that is equipped with a computer that acts the same way of human senses.

3) In the dangerous cities or counties the "Car X" will be sent to collect information that is useful for the military or the FBI and for the national security.

The Explanations of the Flow Chart for Car X:

1) See-Camera-in and out the wall-check for weapon: "Researchers at Massachusetts Institute of Technology have developed the 3D motion tracker which uses radio waves to mark a person's location. The device can plot moves and even follow gestures, meaning that a person can switch a light off in a separate room just by raising their arm." Read more: http://www.dailymail.co.uk/news/article-2523866/Motion-tracking-technology-walls-WITHOUT-camera.html#ixzz3cBfsNncl

   Follow us: @MailOnline on Twitter | DailyMail on Facebook

2) Hear-listen and distinguish language-translate: Researchers at Microsoft have made software that can learn the sound of your voice, and then use it to speak a language that you don't. The system could be used to make language tutoring software more personal, or to make tools for travelers. http://www.technologyreview.com/news/427184/software-translates-your-voice-into-another-language/

3) Smell –analyze air: CONTAM is a multizone indoor air quality and ventilation analysis computer program designed to help you determine: (a) airflows: infiltration, exfiltration, and room-to-room airflows in building systems driven by mechanical means, wind pressures acting on the exterior of the building, and buoyancy effects induced by the indoor and outdoor air temperature difference. (b) contaminant concentrations: the dispersal of airborne contaminants transported by these airflows; transformed by a variety of processes including chemical and radio-chemical transformation, adsorption and desorption to building materials, filtration, and deposition to building surfaces, etc.; and generated by a variety of source

mechanisms, and/or (c) personal exposure: the predictions of exposure of occupants to airborne contaminants for eventual risk assessment. http://www.nist.gov/el/building_environment/contam_software.cfm

4) Taste-analyze air or substance food/dynamite: At MIT, Professor George Benedek had a research to study the differences between energy of food and dynamite. "a Milky Way® candy bar contains more energy than a stick of dynamite. The candy bar contains 200 food Calories. That's 200,000 physicist calories or about 840,000 joules! Nearly a megajoule! A megajoule of energy from a candybar can perform enough work to lift an average 70-kilogram human being 1200 meters in the air. That's higher than the cliff face of Yosemite's El Capitan. No stick of dynamite can do that! In fact, an ounce of dynamite produces only one-quarter as many calories when it explodes as an ounce of sugar does when it burns." http://isaac.exploratorium.edu/~pauld/activities/food/countingcalories.html

5) Touch-analyze surface –in and out: "X-ray photoelectron spectroscopy is a widely used method of determining the chemical composition of a surface." University of Maryland created surface analysis center and here is the address: **Surface Analysis Center,** XPS Analysis Request Form, Department of Chemistry & Biochemistry, University of Maryland, College Park 20742, (301) 405-4999. X-ray photoelectron spectroscopy is a widely used method of determining the chemical composition of a surface. http://www.chem.umd.edu/sharedinstrumentation/surface-analysis-center/

6) Danger: When the Car X tested the environment and all the senses were positive the Car X computer will call the FBI.

7) Car X will move to another location and repeat the above procedures.

CHANGING CLIMATE AND IMMINENT DROUGHT

COPYRIGHT 2015

CHANGING CLIMATE AND IMMINENT DROUGHT

The Problem:

1) Throughout the history of humans on earth the changing climate from agriculture land and raining to desert and drought is done with unknown reasons or with uncontrollable reasons.

2) The changing climate from agriculture land and raining to desert and drought is done naturally and mostly without the interference of man-made pollutions.

3) The Sahara Desert in Africa and the 90% desert of Egyptian land were occurred in the history without the interventions of humans.

4) Burning the forest's trees in the west coast of the United States and the drought are not happening in the African forest in Central Africa.

5) There are two possible reasons for burning the trees in the west coast of the USA. The first reason probably it is a result of domestic terrorists. And the second reason probably it is a result of natural reasons and changing climate on earth or the ever changing position of the Solar System.

6) The drinking water for these reasons will be disappearing from the west coast and other parts of the world.

Observations:

1) The drinking water is needed in the submarines and the NAVY ships.
2) The drinking water will be needed around the world due to changing the climate naturally or by the contribution of man-made pollutions.

The Solution:

The Strategies are:

1) The Saudi Arabia is converting the sea water to drinking water for many years.
2) Converting the sea water to drinking water in the Saudi Arabia is expensive.
3) Removing the salt from the sea water is possible.
4) **Reverse osmosis:** Scientists expose the salt water to increasing pressure, which forces the water to move through a membrane. The membrane allows water molecules to move through it while blocking the salts. Like a filter, the membrane removes the salt and produces fresh water. This figure shows what this process looks like. http://www.dummies.com/how-to/content/environmental-science-how-to-create-fresh-water.html

Credit: Illustration by Wiley, Composition Services Graphics

5) Membrane can be replaced or rehabilitated/restored mechanically to a new useable condition.
6) Membrane processes use semi-permeable membranes and pressure to separate salts from water.
7) AMPAC company is well known for its **reverse osmosis plant** water purification systems. http://www.ampac1.com/?gclid=CLmals-FkcgCFcQUHwodxysLfQ
8) **Natural osmosis:** is a solution that is less concentrated will have a natural tendency to migrate to a solution with a higher concentration.

http://puretecwater.com/what-is-reverse-osmosis.
html#understanding-reverse-osmosis

9) **Reverse osmosis:** <u>Reverse Osmosis is the process of Osmosis in reverse</u>. Whereas Osmosis occurs naturally without energy required, to reverse the process of osmosis you need to apply energy to the more saline solution. A reverse osmosis membrane is a semi-permeable membrane that allows the passage of water molecules but not the majority of dissolved salts, organics, bacteria and pyrogens. However, you need to 'push' the water through the reverse osmosis membrane by applying pressure that is greater than the naturally occurring osmotic pressure in order to desalinate (demineralize or deionize) water in the process, allowing pure water through while holding back a majority of contaminants. http://puretecwater.com/what-is-reverse-osmosis.html#understanding-reverse-osmosis

COPY KEY NEXT TECHNOLOGY

COPYRIGHT 2015

COPY KEY NEXT TECHNOLOGY

While the researcher is reading an article to support her/his research the need to copy multiple quotes from one/multiple application(s) to other application(s) is great. The following is a computer programming flowchart to illustrate my idea of the program sequences, for the computer industry, to be able to add this technology to the computer software/hardware.

COPY KEY NEXT TECHNOLOGY

FLOWCHART

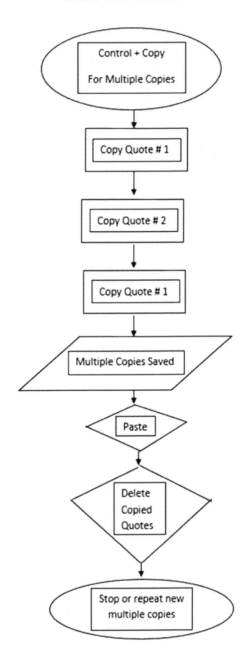

DEFENSE ALTERNATIVES FOR NAVY SHIPS

COPYRIGHT 2016

DEFENSE ALTERNATIVES FOR NAVY SHIPS

The Problem:

For sinking Navy Ships the struggle for the ship crew members is deadly. USS Indianapolis is the example of the devastating 4 days which the crew members were eating by the sharks.

The Idea:

The first alternative is to combine the submarine advantage with the destroyers or any other type of the Navy ships. The second alternative is to detect the enemy torpedoes by a high frequency radar.

The First Solution:

1) The Navy Ships must be able to resist the sudden sinking as a result of the enemy torpedo attacks.
2) Combining submarine/destroyer structures must be done to prevent the damaged ship from sinking.
3) Individual/separated air cylinders that will be able to float the wreckage of the damaged ship must be installed permanently in the inner structures of all the Navy ships.
4) The air cylinders must be in protective structures throughout the inner cavity of the Navy ship.
5) The crew of the destroyed Navy ship will be able to hold in the wreckage, the worst scenario, until the help arrives.

The Second Solution:

1) By detecting the enemy torpedoes by a radar which is operated by a high frequency radio the Navy ships will be able to fire anti-torpedoes guided missiles ship-to- torpedoes against the enemy torpedoes.

2) Because the radar's waves are high frequency waves then transmitting the high frequency in the distance of water to detect the enemy torpedoes will not be possible.

3) Locating/placing the radar equipment to detect the enemy torpedoes on the four top sides of the Navy ship, that is, on the top front, the top rear, the top right side, and the top left side of the navy ship will be the solution to overcome the inability of transmitting the high frequency in the distance of water.

4) The radar will be locating the enemy torpedoes and the speed.

5) The enemy torpedoes travel a few inches from the ocean surface.

6) As usual the high frequency radar will be used from the top of the Navy ship aimed toward the possible enemy torpedo attacks to detect them.

7) Anti-torpedo guided missiles ship-to-enemy torpedoes will be equipped to locate the enemy torpedoes according to the high frequency radar's instructions.

DRONES INCEPTION AND DESTINATION-PART 1&2

COPYRIGHT 2013-2015

DRONES INCEPTION AND DESTINATION-PART 1

The Problem:

Delivering the drones to the enemy activity zone is slow and not practical. The drone is limited to taking photos, spying, and carrying missiles.

The Solution:

The solution is to combine a rocket and shuttle in one flying vehicle. This improved flying vehicle is equipped to fly in the atmosphere and to orbit the earth in space. The reason this flying vehicle should be able to orbit the earth is to shorten the time to deliver the drones.

The Idea:

1) To create an improved flying vehicle (IFV) to carry the drones to the enemy's location activities whether we are at war or not.
2) The drones will be designed with folded wings to fit the improved flying vehicle.
3) The Improved Flying Vehicle (IFV) will be designed to carry inside the vehicle from 5 to 10 drones or any suitable number of drones.
4) When the drones are released over the war zone the folded wings will expand.
5) The IFV will be launched from any military base around the globe regardless of the distance between the military base

and the war zone because the IFV will be launched to the earth orbit in a few minutes and reached/descended over the war zone in a few minutes.

6) The IFV will be operated by a pilot(s) who will drive the IFV away from the war zone faster than the speed of sound and after releasing of the drones is completed.

7) The IFV will return to the nearest military base.

8) The IFV must release the drones in an extreme high altitude that is out of reach of the antiaircraft missiles.

9) Two options are remained for the IFV which are to return to the orbit or to fly away from the war zone by exceeding the speed of sound.

10) If the military base is not approachable the drone itself will be used as a guided missile to hit a target or it will be self destructed after the mission is completed.

DRONES INCEPTION AND DESTINATION-PART 2

"Membrane Optic Imager Real-Time Exploitation (MOIRE). To meet national security requirements around the world, it would be optimal to have real-time images and video of any place on earth at any time—a capability that doesn't currently exist. Today, aircraft are used to meet much of the military's imagery requirements. However, because of the quantity of aircraft needed, and because aircraft do not fly high enough to see into denied territories, spacecraft also play a vital role in providing the imagery data needed for successful military planning and operations" (TTO Programs).

1) In the war time neither infantry troops nor the aircraft should be used in today's war as vehicle for winning a war or to obtain real-time images and videos. The military drones should be used. If we know the military drone last 36 hours flying we can have an improved flying vehicle (IFV) to carry the drones, 5 to 10 drones, to the enemy's location activities.
2) The drones which are released over the enemy location in high altitude will circulate the enemy geographic location for 36 hours to take real-time images and videos.
3) The images will locate the on ground enemy troops.
4) The enemy troops will not be able to remain in the same hiding place for 36 hours.
5) If the enemy troops were able to hide for 36 hours the first wave of drones will be replaced by other wave of drones that will begin the real-time images and videos.

6) When the first wave of drones are about to leave the enemy geographic location after about 36 hours the second wave will join the first wave to make sure our military did not miss the enemy movement and relocation.

7) The above steps should be repeated until the enemy is destroyed by those drones which are equipped with weapon.

EARLY DETECTIONS FOR QUIET SUBMARINES

COPYRIGHT 2013 - 2015

EARLY DETECTIONS FOR QUIET SUBMARINES

The Problem:

Delivering the unmanned vessel to the enemy activity zone is slow and not practical. The quiet submarines are not detected in the deep water.

The Solution:

The *first solution* is to map the oceans and the seas with early detection devices. The devices consist of two sensors. One will be floating near the surface of the water and it will be aimed to the sensor that is floating a few feet above the ocean floor and the second sensor will be floating two or three feet from the ocean floor and will be aimed to the first sensor that is floating under the surface of the ocean. The two sensors will be capable of detecting the location of the submarine that is passing by between the two sensors. The sensors will produce radio waves which are aimed to the satellite that will communicate the coded signal to the Navy.

The *second solution* is to launch geographical device to orbit the earth and locate all the submarines in the oceans. This technology is already used for geographical purposes. Taking inventory for all submarines which are in the oceans and the seas is important for protecting this country.

The First Idea:

1) The signal of the sensors is coded to position their location.
2) The coded signals should not interrupt the computers aboard the enemy's submarine.
3) The coded signals will be transmitted to the satellite which will send the signals to the Navy.
4) The Navy will decode the signal and locate the enemy quiet submarine on the ocean map.
5) The sensors will be powered by super batteries.
6) The sensors will be randomly distributed in the hot spots but their locations will be known on the ocean map.
7) The sensors must be able to code the size of the submarine.
8) The power of the sensors must cover a few miles.
9) Two options are remained. The enemy submarine will be destroyed by the Air Force warfighters or by the unmanned vessels which are carried on the Improved Flying Vehicle if a military base is not approachable.
10) This Improved Flying Vehicle (IFV) is equipped to fly in the atmosphere and to orbit the earth in space. The reason to create and use this flying vehicle because it should be able to orbit the earth is to shorten the time to deliver the unmanned vessel.
11) The IFV is a combination of rocket and shuttle.
12) The IFV will be launched from any military base around the globe regardless of the distance between the military base and the enemy's submarine because the IFV will be launched to the earth orbit in a few minutes and reached/descended over the enemy's submarine in a few minutes.
13) The IFV will be operated by a pilot(s) who will drive the IFV away from the war zone or the enemy's submarine faster than the speed of sound and after releasing of the unmanned vessels is completed.
14) The IFV must release the unmanned vessels in an extreme high altitude that is out of reach of the antiaircraft missiles.

The unmanned vessels will be released and lowered in the atmosphere by parachutes.

15) The parachutes will be separated immediately from the unmanned vessel upon landing on the water.

16) The unmanned vessel's remote navigator will be directed to follow, detect, or destroy the enemy quiet submarine.

17) The IFV could have one or more unmanned vessels.

The Second Idea:

1) The geological device in the orbit will allow the Navy to locate all the submarines which are in the oceans and the seas either quiet submarines or not quiet submarines.

2) The location of each submarine will be known to the Navy.

3) The quiet submarine will be distinguished and detected because we already know the location of the noisy submarines.

4) Two options are remained. The enemy, quiet or not quiet, submarine will be destroyed by the Air Force warfighters or by the unmanned vessels which are carried on the Improved Flying Vehicle if a military base is not approachable.

5) This Improved Flying Vehicle (IFV) is equipped to fly in the atmosphere and to orbit the earth in space. The reason to create and use this flying vehicle because it should be able to orbit the earth is to shorten the time to deliver the unmanned vessel.

6) The IFV is a combination of rocket and shuttle.

7) The IFV will be launched from any military base around the globe regardless of the distance between the military base and the enemy's submarine because the IFV will be launched to the earth orbit in a few minutes and reached/descended over the enemy's submarine in a few minutes.

8) The IFV will be operated by a pilot(s) who will drive the IFV away from the war zone or the enemy's submarine faster than the speed of sound and after releasing of the unmanned vessels is completed.

9) The IFV must release the unmanned vessels in an extreme high altitude that is out of reach of the antiaircraft missiles. The unmanned vessels will be released and lowered in the atmosphere by parachutes.

10) The parachutes will be separated immediately from the unmanned vessel upon landing on the water.

11) The unmanned vessel's remote navigator will be directed to follow, detect, or destroy the enemy quiet submarine.

12) The IFV could have one or more unmanned vessels.

EGYPT AND NATO THE STRATEGY

COPYRIGHT 2014

"Dear President Obama

I am a Coptic Christian originally from Egypt. I know El Qaida/ISIL terrorists kill Christians as well as innocent Muslims. I was born in El Giza-Egypt and I lived in Cairo for 32 years. All my best friends were Muslims. I loved them all. I really want you to know that Ancient Egyptians as well as the modern Egyptians are great and compassionate people. We love Egypt. What you do not know is that Egyptians love the United States of America, too. I am an American citizen. I ask you to review the relationship between the USA and Egypt and between the North Atlantic Treaty Organization (NATO) and Egypt. I ask you to accept Egypt to the NATO's nations to be protected and, on the other hand, defending the principles and the political systems of NATO's members. Egyptian people are smart and honest people. When Egypt becomes a member of NATO the Islamic world will deepen the relationship with the west. For Egypt to be a member of NATO, the new status of Egypt will demolish the existence of terrorists around the world. The struggle of power around the world particularly between the USA and Islamic world will be balanced. The Egyptian military prevented Egypt from the dictatorship of the fanatic/terrorist Muslims/Muslim Brothers. The ordinary Egyptian Muslims, Egyptian Christians, and Egyptian Jewish want to live in peace and prosper. Please, Mr. President Obama, help the USA gain power and explore a new world." **(Milad, 2014)**

12/30/2013

ELECTRIC CARS THE IMPROVED NEXT GENERATIONS

410-282-8708
COPYRIGHT 2013

ELECTRIC CARS THE IMPROVED NEXT GENERATIONS

The Problem: 1) The first problem with the electric cars is the ordinary rechargeable batteries. 2) The second problem is the unpractical recharged system which will consumes hours for recharging the batteries. 3) The third problem is the power stations and how the owner of the electric car will need to be alert when the batteries need to be recharged and how far from the next station. 4) The fourth problem is that the costs of the cars and the batteries are not cost effective.

The solution: 1) The solution is to creat a super portable rechargeable battery which will be replaceable by a spare super portable rechargeable battery. 2) The super uncharged battery will be recharged at home or at a station if the owner of the electric cars prefers. 3) The ordinary rechargeable batteries will be receiving and recharged their power from the super portable rechargeable battery. 4) The owner of the electric car may own more than one spare super rechargeable battery. 5) In the electric car, the controller will receive the power via the ordinary rechargeable battery to the wire loop to produce the magnetic field that transfers the energy to the shaft.

The New Market: Creating a super portable rechargeable battery will create a new practical product that will help the owners of the electric cars to carry in her/his vehicle a spare super portable rechargeable battery.

The Idea: The idea is to create a super portable/replaceable rechargeable battery that will have the capability to be recharged from home electricity and to recharge the electric cars' ordinary batteries. In other words, the super portable rechargeable battery will be recharged to recharge the ordinary batteries. The super portable rechargeable battery will be recharging the ordinary batteries continuously. The super portable rechargeable battery will keep the level of the power needed steady in the ordinary batteries to produce the magnetic field that transfers the energy to the shaft.

FUTURISTIC CRIPPLING NETWORK

COPYRIGHT 2015

FUTURISTIC CRIPPLING NETWORK

The Problem:

1) Today China announced the elimination of the ground troops in its military https://www.washingtonpost.com/world/asia_pacific/mixing-doves-and-heavy-weapons-china-marks-world-war-ii-anniversary/2015/09/02/1b046008-5028-11e5-b225-90edbd49f362_story.html.
2) China will be building sea and air forces to modernize its military.
3) Relying on the orbit satellites networking is a weak strategy.

The Solution:

The Strategies are:

1) In the beginning of a war, the first strategy is to aim high frequency electromagnetic signals against the enemy's satellites to create electrical interference to the enemy's defense electronic equipment to cripple its sea and air military forces.
2) As another alternative, a specialized satellite that could be specialized to produce high frequency electromagnetic signals will be orbiting the earth and targeting the enemy's satellites.
3) Also the high frequency electromagnetic signals could be used against the enemy's military bases to create electrical interference to the enemy's guided missiles electronic system.

4) In the sea and ocean, creating small quiet submarines that are specialized to produce high frequency electromagnetic signals could be the solution to cripple the enemy's quiet submarines when creating electrical interference in the enemy's quiet submarines.

5) Exposing the quiet submarines should be done when launching geographical device to orbit the earth and locate all the submarines in the oceans. This technology is already used for geographical purposes. Taking inventory for all submarines which are in the oceans and the seas is important for protecting this country. The NAVY will launch the satellite archaeology "Some 400 miles up in space, satellites collect images that are used to identify buried landscapes with astonishing precision." (Bloch, 2013). "University of Alabama at Birmingham archaeologist Sarah Parcak, is a pioneer in using satellite imagery in Egypt." (Bloch). The satellite archaeology is not limited to buried landscapes.

6) Creating satellites and small submarines that are equipped/specialized to produce high frequency electromagnetic signals could be the solution to cripple the enemy's quiet submarines, guided missiles, air-to-air missiles and ground-to-air missiles when creating electrical interference in the enemy's electronic offense/defense systems and its satellites networking.

7) Creating our warplanes that are independent from the satellites networking is the most crucial invention/demand to defeat the enemy which is relying on the satellites networking.

8) Our warplanes must be equipped with self-efficient-networking that is not connected with the orbit satellites networking.

9) This research's predicted/suggested networking could be ground satellites that are located on the ground military bases and they do not orbit the earth.

Reference

Bloch, Hannah (2013) The New Age of Exploration. Satellite Archaeology. Retrieved November 29, 2014 from http://ngm. nationalgeographic.com/2013/02/125-explore/satellite-archaeology

HYBRID AIRCRAFT

COPYRIGHT 2015

HYBRID AIRCRAFT

The Problem:

"Vertical Takeoff and Landing Experimental Plane (VTOL X-Plane): VTOL X-Plane challenges industry and innovative engineers to create a single hybrid aircraft that would concurrently push the envelope in four areas:

o   Speed: Achieve a top sustained flight speed of 300 kt-400 kt
o   Hover efficiency: Raise hover efficiency from 60 percent to at least 75 percent
o   Cruise efficiency: Present a more favorable cruise lift-to-drag ratio of at least 10, up from 5-6
o   Useful load capacity: Maintain the ability to perform useful work by carrying a useful load of at least 40 percent of the vehicle's projected gross weight of 10,000-12,000 pounds." (TTO Programs).

The Solution:

1) To achieve a top sustained flight speed of 300 kt-400 kt the helicopter will be equipped with jet engines as shown in the design that is located in page 3.
2) To raise hover efficiency from 60 percent to at least 75 percent the helicopter horizontal propeller will remain intact with the flexibility to be folded during the jet engines maneuver.
3) The new design of the Hybrid Aircraft will present a more favorable cruise lift-to-drag ratio of at least 10, up from 5-6

because the helicopter will be equipped by wings for the aerodynamic mobility.

4) No change in the useful load capacity of the helicopter as a Hybrid Aircraft and it will maintain the ability to perform useful work by carrying a useful load of at least 40 percent of the vehicle's projected gross weight of 10,000-12,000 pounds

5) The Hybrid Aircraft will be converted to a jet engines aircraft during the jet engines maneuver.

6) The tank gears could be maintained or removed. The tank gears originally were designed to be made of light weight aluminum for uneven ground.

7) The design in page three lacks the wings which are necessary during the jet engines maneuver.

8) The design in page three is equipped by the jet engines on the top of the helicopter and under the helicopter horizontal propeller.

9) The design in page three must have wings which are necessary during the jet engines maneuver.

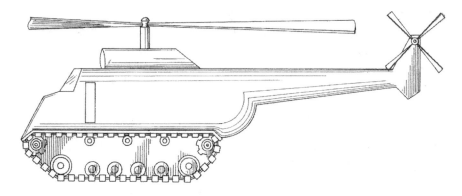

United States Patent

[19] Milad

[11] Patent Number: Des. 292,193 [45] Date of Patent: ** Oct. 6, 1987

[54] COMBINED HELICOPTER AND TANK

[76] Inventor: Anis I. Milad, 29-38 Yorkway, Baltimore, Md. 21222

[**] Term: 14 Years

[21] Appl. No.: 690,587

[22] Filed: Jan. 11, 1985

[52] U.S. Cl.............................D12/2; *D21190;*
D211l31

[58] Field of Search.......*D211131,* 85; D12/214, D12/12; 446/230,37,237,433,465; 244/1 R, 2, 17.11, 17.17

[56] References Cited

U.S. PATENT DOCUMENTS

D. 135,849 6/1943 Gibson..................D21/131
D.138,329 7/1944 Wasson..................D21/85
D.169,229 3/1953 Tober....................D21/131
D. 284,594 7/1986 Fisher....................D21/85
D.287,378 12/1986 Ohno..................D21/131

1,788,440 l/t931 Prell.
2,019,482 11/1935 Barnes.
2,424,769 7/1947 Page.
2,560,747 7/1951 Sievers.
3,761,040 9/1973 Cummins.

Primary Examiner-Charles A. Rademaker
Attorney, Agent, or Firm-Leonard Bloom

[57] CLAIM

The ornamental design for a combined helicopter and tank, as shown and described.

DESCRIPTION

FIG. 1 is a side elevational view of a combined helicopter and tank showing my new design, the side opposite being substantially a mirror image;

FIG. 2 is a top plan view thereof;

FIG. 3 is a bottom plan view thereof;

FIG. 4 is a front elevational view thereof; and,

FIG. 5 is a rear elevational view thereof. Portions of the blade members have been shown In fragment im FIGS. 2 and 3 for ease of illustration.

ILLEGAL DRUGS THE SOLUTION

COPYRIGHT 2016

ILLEGAL DRUGS THE SOLUTION

The Problem:

The illegal drugs are spreading in the United States of America and around the world. The illegal drugs are the imminent reasons for destroying the existence of the current civilizations. The USA and the Western civilizations are targeted. In Afghanistan and Mexico there are thousands of acres which are planted by the illegal drugs.

The Solution:

The military Air Force must have an important role to locate/destroy the illegal drugs farmland by using pesticide Air Force airplanes to spray poison chemical to kill the illegal drugs' tree, tree root, and plants. For example, we are already occupying Afghanistan and eliminating the farms of the illegal drugs are in open land and depends on the rain for growth. Another example, The drug cartels in Mexico also grow the illegal drugs in open farmland. If the illegal drugs farm is under a dome then the dome must be destroyed.

The following is a practical approach for eliminating the illegal drugs and causing a financial disaster to the drug dealers/cartels:

"Soil Surface" "Herbicides applied directly to the soil surface are carried down into the root system with rainfall or watering. The chemicals kill the roots when they come into direct contact with them. Chemicals that work in this way include bromacil, hexazinone and tebuthiuron." http://homeguides.sfgate.com/chemicals-kill-tree-roots-29528.html

In Mexico planting illegal drugs are encouraged by the Mexican government to compete with the flooded Mexican market by the USA U.S. agricultural products. Here are the reasons: "When Mexico and the United States were entering a landmark free trade agreement 16 years ago, one thing was clear: Mexican farmers would initially find it difficult to compete with heavily subsidized U.S. agricultural products."

"The solution: Mexico created a special fund to dole out cash to the poorest and smallest farmers."

"Somewhere along the way, something went wrong. Today, the fund – far from helping the neediest – is providing large financial subsidies to the families of notorious drug traffickers and several senior government officials, including the agriculture minister."

"Revelations of how and to whom the money is being distributed have led to a spasm of demands from legislators to change the system. But, as with most examples of colossal corruption in Mexico, it is unlikely that the program will be overhauled." http://articles.latimes. com/2010/mar/07/world/la-fg-mexico-farm-subsidies7-2010mar07

From the above quote our Air Force must eliminate the illegal drugs from the source the that is the illegal drugs farmlands.

Let's begin with Afghanistan as an experiment.

Arresting the illegal drug dealers and untouch their source of income and wealth, that is the illegal drugs farmland, is not a logical solution. Destroying the illegal drug farmland must be the priority to protect the American people and the American civilization.

ISIL THE FAR RIGHT AND THE MILITARY SOLUTION

COPYRIGHT 2014

ISIL THE FAR RIGHT AND THE MILITARY SOLUTION

ISIL extremists are not terrorists/evil doers but they are, according to our democratic system, Muslim and the far right of the Republican Party.

ISIS (ISIL) extremists are also Sonny. On the other hand, the president of Syria, Bashar el Assad is far left/socialist/Shia/Alawite Muslim.

In Syria Americans are the enemy of Bashar el Assad (Shia) and ISIS (Sonny). The USA should not interfere in this war militarily but we can negotiate with Bashar el Assad to fight ISIS and the reward is to acknowledge his role in the Middle East to gain back the control of Syria.

The crisis in Iraq and Syria began by hanging Saddam Hosin the former dictator of Iraq.

George W. Bush was the reason for the rise/grow of the Muslim Extremists in Iraq and Syria.

Seeking democracy does not have to end with a democratic system but seeking democracy probably will end with a terrifying dictatorship which we are seeing now in Iraq and Syria.

The military should not send the American soldiers to Iraq and Syria because Bashar el Assad can defeat ISIS for USA.

The president should negotiate with Bashar el Assad and supply him with the military equipment he will need to defeat ISIS.

USA accuses Bashar el Assad by killing a few hundreds of his own people and the ironic facts are that George W. Bush was the direct reason for the war in Iraq and Afghanistan and the loss of the lives of 10,000 American soldiers. Comparison was needed here!!!!!

Bashar el assad will purchase the military equipment and we in America will collect the cash.

For us it is a win-win situation. This war in the Middle East is between the Muslim Extremists of the Stone Age and the moderate Muslims.

Let the Muslims reconcile their differences with each other.

If Mohamed the prophet comes in our time he will be shocked to see the Muslim extremists beheaded human beings.

One important thing I remember about Mohamed the prophet is that his wife Kadiga was a business woman who traveled the old world to do business and to be wealthy.

Mohamed the prophet will be shocked again to see the Muslim extremists humiliating the Muslim women in this century and killing innocent Muslim people!!!!!!!!!!!!!

9/1/14 – Updated 9/2/2014

Reference: Milad, A I (2014) *Letter to President Obama*

LANDMINE REMOVAL

COPYRIGHT 2015

LANDMINE REMOVAL

The Problem:

"Difficult terrain and threats such as ambushes and Improvised Explosive Devices (IEDs) can make ground-based transportation to and from the front line a dangerous challenge. Helicopters can easily bypass those problems but present logistical challenges of their own, and can subject flight crew to different types of threats. They are also expensive to operate, and the supply of available helicopters cannot always meet the demand for their services, which cover diverse operational needs including resupply, fire-team insertion and extraction, and casualty evacuation."

Unpractical Solution:

"The flare is safe to handle and easy to use. People working to deactivate the mines – usually members of a military or humanitarian organization – simply place the flare next to the uncovered land mine and ignite it from a safe distance using a battery-triggered electric match. The flare burns a hole in the

land mine's case and ignites its explosive contents. The explosive burns away, disabling the mine and rendering it harmless." http://www.nasa.gov/home/hqnews/1999/99-129.txt. It is an unpractical solution because the landmines are located in unknown locations.

The Solution of this research:

The components of this research are:

1) It is entirely possible to create static electricity, and even lightning using this method. Van de Graaf generators, for example, use rubbing to generate voltages in excess of a 1,000,000V. However, it's a very inefficient method for generating power. Dynamo generators (the standard generator) are surprisingly efficient. http://www.askamathematician.com/2010/05/q-would-it-be-possible-to-generate-power-from-artificial-lightning/

2) **Fuse is a component of the landmine. Fuse is a flammable material which is defined as following: "Fuse** - A combustible material used to ignite an explosive charge. http://science.howstuffworks.com/landmine1.htm.

The Strategy:

1) Creating a high voltage artificial lightning using dynamo.
2) Installing the dynamo in a helicopter which will be used for this purpose only.
3) While the artificial lightning is produced, the helicopter will be flying in a safe altitude over the landmine field. There is no need to locate the landmines.
4) The artificial lighting will trigger/ignite the combustible material which will ignite the **detonator which will ignite the** larger/remaining amounts of explosive.
5) The artificial lightning will continue to be produced and the explosions of the landmines will continue. The helicopter could repeat theses procedures to be sure all the landmines are exploded in the intended zone.
6) During the war, the helicopter could create a quick safe road which can make the ground-based transportation to and from the front line a safe operation.

LIGHTNING DETECTION

COPYRIGHT 2015

LIGHTNING DETECTION

The Problem:

1) The inability to detect/target specific live biological terrorists in the war zone.
2) The inability to terminate huge number of the biological terrorists in a matter of minutes in the open war zone.
3) The inability to advance with the ally's war vehicles in the enemy's war zone because our inability to detect/destroy the anti-tank landmines and the suicidal bombers.
4) The inability to detect/destroy the anti-personnel landmines in the cities.

The Solution:

The Strategies are:

5) The first strategy is to equip our warplanes, helicopters or drones, with dynamo generators that able to produce in excess of 1 million volts.
6) The purpose of the dynamo generators and the excess of 1 million volts are to produce the artificial lightning which will be aimed toward the ground to eliminate the biological terrorists in the war zone.
7) Creating helicopters or drones that are equipped/specialized to produce high/deadly voltages could be the solution to cripple the enemy's specific live biological terrorists in the war zone, to terminate huge number of the biological terrorists in a matter

of minutes in the open war zone, to be able to detect/destroy the anti-tank landmines and the suicidal bombers, and to be able to detect/destroy the anti-personnel landmines in the cities.

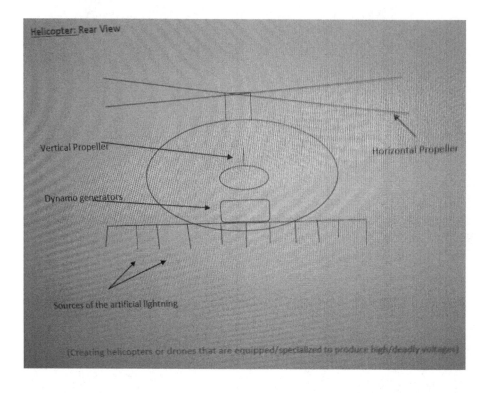

Helicopter: Rear View

Vertical Propeller

Horizontal Propeller

Dynamo generators

Sources of the artificial lightning

(Creating helicopters or drones that are equipped/specialized to produce high/deadly voltages)

MACH 12 WARPLANE

COPYRIGHT 2014

MACH 12 WARPLANE

The first problem that is associated with the fighter jet in general is the limited capability of the pilot as a human being and the speed of the fighter jet. The second problem that is associated with the fighter jet is the maximum speed of the fighter jet and the huge distance between the American military bases and the location of the enemy. The third problem that is associated with the fighter jet is the slow response to the satellite signals (pictures) which locate a particular location of the military equipment of the enemy because the slow response is due to the huge distance between the USA military bases and the location of the enemy.

The main concern is to narrow the response time between the signals of the satellite and destroying the enemy. The speed of F-35 is Mach 1.6 but the speed of the rocket is Mach 10 (China is experimenting with Mach 12). When producing the rocket in the same way the F-35 is produced to be able to carry several bombs/guided missiles/clusters, intelligent data, stealth, and satellite networking connections the delivery of the bombs for several enemy locations will be faster and instant. The suggested warplane rocket will be unmanned so the physical pilot factor will be eliminated. As an alternative, the warplane rocket will be operated remotely via the satellite. As another alternative, the warplane rocket will be operated by a pilot who will be wearing a space suit.

Analyses:

1 Mach [Dry Air @ 273 Kelvin] = 741.81729 miles/hour

If the speed of the rocket is Mack 10 and it is launched from Washington DC, the distance from Washington DC to Damascus = 5877 miles

10 Mach [Dry Air @ 273 Kelvin] = 7,418.1729 miles/hour 5877 / 7,418.1729 = .7922 hour = 60 * .79 = 47 minutes

China is experimenting on Mach 12 rocket, so If the speed of the rocket is Mack 12 and it is launched from Washington DC, the distance from Washington DC to Damascus = 5877 miles.

12 Mach [Dry Air @ 273 Kelvin] = 8,901.80748 miles/hour 5877 / 8,901.80748 = .6602 hour = 60 * .66 = 39.60 minutes

If the speed of the rocket is Mack 10 and it is launched from the United Kingdom - Scotland - Edinburgh, the distance from Scotland – Edinburgh to Damascus = 2433 miles

10 Mach [Dry Air @ 273 Kelvin] = 7,418.1729 miles/hour 2433 / 7,418.1729 = .3279 hour = 60 * .32 = 19.2 minutes

To utilize Mach 12 rocket:

12 Mach [Dry Air @ 273 Kelvin] = 8,901.80748 miles/hour 2433 / 8,901.80741 = .27 hour = 60 * .27 = 16.2 minutes

When the rocket, Mach 12, is launched from the United Kingdom - Scotland – Edinburgh, the rocket will take 16.2 minutes from receiving the satellite (pictures) signals of the enemy's convoy (trucks and tanks) in Damascus to the moment of destroying those trucks and tanks.

In comparison, F-35 that is launched from the United Kingdom - Scotland – Edinburgh will take 2.04 hours to reach Damascus.

1.6 Mach [Dry Air @ 273 Kelvin] = 1,186.907664 miles/hour

2433 / 1,186.907664 = 2.04 hours =60 * 200.04% = 120.24 minutes

References

Milad, A I (2014): Letters to the President as of September 3, 2014: Copyrighted 2014

MILITARY MODERN RIFLES

COPYRIGHT 2013

MILITARY MODERN RIFLES

The Problem:

"For military snipers, acquiring moving targets in unfavorable conditions, such as high winds and dusty terrain commonly found in Afghanistan, is extremely challenging with current technology. It is critical that snipers be able to engage targets faster, and with better accuracy, since any shot that doesn't hit a target also risks the safety of troops by indicating their presence and potentially exposing their location." (TTO Programs)

The Solution:

The solution is to combine in the rifle a telescope, remote control, and battery; and to combine in the bullet an operational fin and sensor.

The Idea:

1. The bullet will be designed with a fin.
2. The fin must be pushed down in the groove; then the sniper will load the bullet.
3. Immediately when the sniper shoots, and the bullet is released toward the target, the bullet internal fin will be raised.
4. The vertical fin of the bullet is connected and controlled by a sensor which is attached to the bullet.
5. The sensor will be receiving the directions from the signal of the remote control that is attached to the rifle.

6. When the bullet changes its position to the right or the left the remote control that is connected to rifle will adjust the sensor/fin to the right or the left according to the continuously emitted signal of the remote control of the rifle.

7. When the sniper adjusts the rifle to the right or the left the sensor that is connected to the fin of the bullet will adjust the fin to the right or the left according to the motion of the remote control and the rifle.

8. The real time will take a few seconds from the sniper to the target.

9. The sniper must look through the telescope to follow the target.

10. The bullet will follow the sniper's movement and the signal from the remote control.

11. The sniper must keep looking at the target through the telescope until the bullet hits the target.

12. Experimentations should be conducted before using this idea/system that is described above.

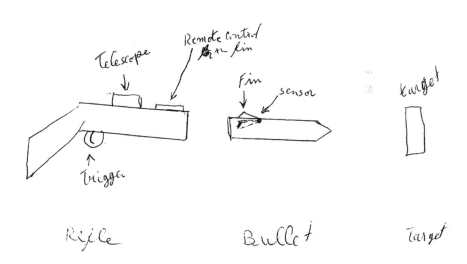

NEGOTIATING WITH THE ENEMY

COPYRIGHT 2016

NEGOTIATING WITH THE ENEMY

The Problem:

The enemy of the United Stated in the present and the future time is an unorganized military and in the Ancient Egypt they were known as "barbarians" who are in a constant attack to the Ancient Civilization. USA is the modern and only civilization in our time and the current barbarians will be in a constant war with the USA. There will be certain destructive holy verses to be adopted by the barbarians to reason their attack to the current civilization.

The Idea:

Negotiating with the barbarians is important to keep the civilization strong and to continue its mission technologically, philosophically, and economically superior and intact.

The Solution:

1) Negotiating with barbarians is an economical approach to win the war with enemy.
2) Around the negotiation table you will be able to coerce your demands. You will be able to demand a peaceful living/ relationship with other religions including Christians.
3) You will be able to excavate/import the crude oil from their geographic locations to cause earthquakes as a result of removing the crude oil from the earth crust.
4) The crude oil naturally is acting as shocker absorbers.

5) You will be able to locate their developed/organized military by the satellite after the enemy government is in control.
6) You will be able to threaten the existence of the enemy's unified country every moment.
7) You will be able to place CIA operatives inside the borders.
8) To say USA nuclear capability is 32,000 nuclear war head it means the USA is dealing with a little skirmish encounter.

May 13, 2016

ON DEMAND SATELLITE IMAGERY

COPYRIGHT 2014

ON DEMAND SATELLITE IMAGERY

The Problem

"Today, the lowest echelon members of the U.S. military deployed in remote overseas locations are unable to obtain on-demand satellite imagery in a timely and persistent manner for pre-mission planning. This is due to lack of satellite overflight opportunities, inability to receive direct satellite downlinks at the tactical level and information flow restrictions." (TTO Programs).

"DARPA's SeeMe program aims to give mobile individual US warfighters access to on-demand, space-based tactical information in remote and beyond- line-of-sight conditions. If successful, SeeMe will provide small squads and individual teams the ability to receive timely imagery of their specific overseas location directly from a small satellite with the press of a button — something that's currently not possible from military or commercial satellites." (TTO Programs).

The Solution

The first solution is to launch satellite archaeology "Some 400 miles up in space, satellites collect images that are used to identify buried landscapes with astonishing precision." (Bloch, 2013). "University of Alabama at Birmingham archaeologist Sarah Parcak, is a pioneer in using satellite imagery in Egypt." (Bloch). The satellite archaeology is not limited to buried landscapes.

The second solution is to launch Space Radar (SR) to orbit the earth. The SR should be equipped to return to the earth for maintenance and to be re-launched/reused. The SR will be equipped to focus on the enemy targets, such as, ground convoy, aircraft, cruise missiles, and surface-to-air missiles and to locate them precisely and instantly. The continuous radar signals that are sent/received instantly from the SR and the targets will be emitted from the satellite radar which will send the signals to the stealth long range guided missile. The signals that are received by the stealth long range guided missile will be updated instantly to locate accurately the enemy targets such as ground convoy, aircraft, cruise missiles, and surface-to-air missiles. As a result, the stealth long range guided missile will change its direction accordingly until it hits the target.

The third solution is to launch Space Radar (SR) to orbit the earth. The SR should be equipped to focus on the enemy targets "to give mobile individual US warfighters access to on-demand, space-based tactical information in remote and beyond- line-of-sight conditions." The warfighters must have a radar screen receiver with Google maps.

Reference

Bloch, Hannah (2013) The New Age of Exploration. Satellite Archaeology. Retrieved November 29, 2014 from http://ngm. nationalgeographic.com/2013/02/125-explore/satellite-archaeology

PERSISTENT SYSTEM FOR MULTIPLE TARGETS

COPYRIGHT 2013

PERSISTENT SYSTEM FOR MULTIPLE TARGETS

The Problem:

"To maintain a decisive tactical advantage in 21st-century combat, warfighters need the ability to safely, rapidly and collaboratively deploy ordnance against elusive mobile targets. Unfortunately, air-ground fire coordination—referred to as Close Air Support or CAS—has changed little since its emergence in World War I. Pilots and dismounted ground agents can focus on only one target at a time and must ensure they hit it using just voice directions and, if they're lucky, a common paper map. It can take up to an hour to confer, get in position and strike—time in which targets can attack first or move out of reach." (TTO Programs).

The Solution:

The components of the solution are: 1) Drone. 2) Radar. 3) Computer software. 4) Location Receivers. 5) Signal connectors. 6) Resources for laser or other weapon. 7) A switch.

The solution is to continue destroying targets as long as the switch is on and the laser or the missiles are not exhausted.

The Idea:

1) The system's illustration is on the following page.
2) The drone, as an alternative, will be equipped with radar.

3) The radar will locate two enemy tanks. The number of tanks could be more than two tanks.

4) The locations of the tanks are sent to the location receivers (LR).

5) The computer software in the drone will select the location receiver LR1.

6) The software will acknowledge the first selection then will select LR2.

7) The selections of the location receivers, LR1and LR2, will be done in a second.

8) The location receivers will send signals to identify the locations of the tanks continuously via the signal connectors to the laser resources, L1 and L2, or any other weapon.

9) The laser will be aimed toward the tanks as long as the switch is on.

10) If the laser is capable to destroy the tanks the time of destroying the tanks must be in seconds or the missiles will be more effective.

11) The information of the multiple moving targets, tanks as examples, will be picked by the radar and the system will continually repeat the operation.

12) The drones could be delivered to the enemy zone by an improved flying vehicle (IFV).

13) The IFV could carry more than 5 drones which will be released in the air.

14) The IFV and the drones will save our warfighters who will operate the drone remotely.

15) The IFV will be operated by a pilot(s) who will drive the IFV away from the war zone faster than the speed of sound and after releasing of the drones is completed.

16) The drones might be programmed in advance to fly over the enemy zone without the human interventions.

Persistent System for Multiple Targets 3

Illustration:

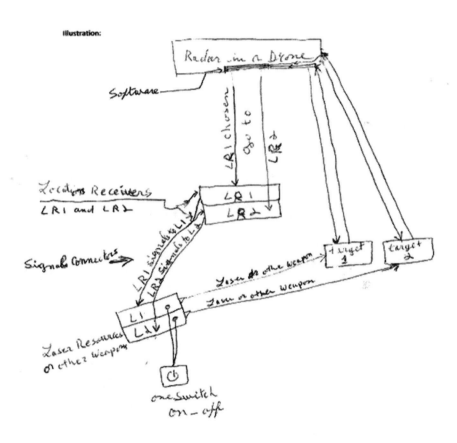

PRIVACY AND UTILITY, THE SOLUTION

COPYRIGHT 2015

PRIVACY AND UTILITY, THE SOLUTION

The Problem:

How can society responsibly reap the benefits of big data while protecting individual privacy?

The Legal Aspects:

The congress and the Supreme Court must classify the privacy into four categories:

1. Confidential Privacy;
2. Explicit Privacy: Volunteerism to share information; and Forced to Inform.

The Ideas:

The Confidential Privacy: The need for confidential privacy is protected by the law. But breaching the privacy is well known fact nowadays. For example, the history of the medical conditions should be kept confidential but recently Johns Hopkins Bay View Hospital in Baltimore Maryland is providing to its patients after the doctor visit a history of their medical conditions which should be kept confidential in the hospital. How the patients will be able to keep their medical records privately is not clear.

The Explicit Privacy: The examples of the explicit privacy, which are categorized as volunteerism to share information, are the social media and blogs; on the other hand, the examples of the category Forced to

Inform are the inventory systems in the department stores, and the drones.

The Solution:

Johns Hopkins Bay View Hospital probably determined it is unpractical to keep the history of the medical conditions confidential. So the Hospital transferred the privacy responsibility to the patients. It is probably legal and acceptable practice. The Supreme Court must define the term, Confidential Privacy, and only after classifying/ categorizing the term, Privacy.

If the law enforcers collected information through drones the damage of the person's privacy is already done and should be evaluated and the violated person should be compensated.

The Big Data:

The Big Data will be programmed. The resources of the information must be the ingredient of the computer program. The computer entry data must be classified the same way the law has been classified. For example if the private information is obtained through drones then the person is forced to inform and by the law s/he must be compensated.

RADAR CRITICAL TERMINATOR

COPYRIGHT 2013 - 2015

RADAR CRITICAL TERMINATOR

The Problem:

"Foreign countries have studied U.S. military actions and are methodically developing strategies and technologies to defeat U.S. Air Dominance. This growing threat is in part attributed to the local numerical advantage many enemy defenses possess." (TTO).

The Solution:

The solution is to launch space radar (SR) to orbit the earth. The SR should be equipped to return to the earth for maintenance and to be re-launched/reused. The SR will be equipped to focus on the enemy target aircraft, cruise missiles, and surface-to-air missiles and to locate them precisely. The continuous radar signals that are sent/received from/to the SR will be emitted to a satellite which will continuously send the signals to the long range guided missile. The signals that are received by the long range guided missile will be updated instantly to locate accurately the enemy target aircraft, cruise missiles, and surface-to-air missiles. As a result, the long range guided missile will change its direction accordingly until it hits the target.

The Idea:

1) The advantage of the SR is to detect anywhere on earth the enemy target aircraft, cruise missiles, and surface-to-air missiles immediately after launch.

2) The long range guided missiles could be on the Navy warships or on the military bases. The guided missile that is about to be launched should be coded to receive the satellite signals.

3) The long range guided missile might be carried internally by 5th generation aircraft (F-22 and F-35), as well as externally on 4th generation aircraft (F-15, F-16, and F-18)

4) The task of the warfighters is to secure/focus the SR on the enemy target(s).

5) Secure/focus the SR could be done without human intervention.

6) The SR automatic intervention could be switched to manual intervention.

7) The purpose of using the SR is to direct the long range guided missiles to hit the enemy target aircraft, cruise missiles, and surface-to-air missiles.

8) The long range guided missiles must be coded and to be read by the SR as friendly missiles. The SR must be equipped to distinguish our missiles from the enemy target aircraft, cruise missiles, and surface-to-air missiles.

9) These long range guided missiles should not be sold to other countries

10) These long range guided missiles could have a self destruction feature that is ignited by the warfighters.

11) The high rank officer might disable the self destruction feature.

12) f the technology allows that the radar and the satellite to be one unit the costs probably will be less than the separate radar and satellite.

13) For stealth enemy missiles, the satellite will be equipped with cameras and videos.

14) Using the Radio waves, television waves, and microwaves also possible means to detect the enemy stealth or not stealth missiles.

Illustration:

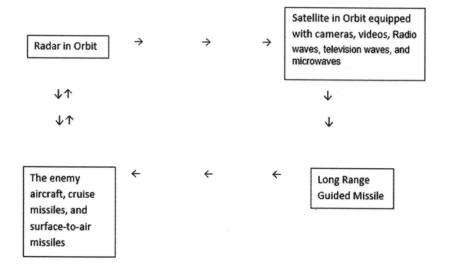

RADAR EXTENSION FOR LONG RANGE MISSILES

COPYRIGHT 2013

RADAR EXTENSION FOR LONG RANGE MISSILES

The Problem:

Anti-ship missiles face a challenge to penetrating sophisticated enemy air defense systems from long range. As a result, warfighters may require multiple missile launches and overhead targeting assets to engage specific enemy warships from beyond the reach of counter-fire systems (TTO).

The Solution:

The *solution* is to launch space radar (SR) to orbit the earth. The SR should be equipped to return to the earth for maintenance and to be re-launched/reused. The SR will be equipped to zoom to the target enemy warship and to locate it precisely. The continuous radar signals that are sent/received to the SR will be emitted to a satellite which will continuously send the signals to the guided missile. The signals that are received by the guided missile will be updated instantly to locate accurately the enemy warship. As a result, the guided missile will change its direction accordingly until it hits the target enemy warship.

The Idea:

1) The guided missiles could be on the Navy warships or on the military bases. The guided missile that is about to be launched should be coded to receive the satellite signals.

2) The task of the warfighters is to secure/zoom the SR on the enemy warship.
3) The purpose of using the SR is to eliminate the direct contact of the warfighters with the enemy warships.
4) If the technology allows that the radar and the satellite to be one unit the costs probably will be less than the separate radar and satellite.

Illustration:

RADIOACTIVE WASTE THE ALTERNATIVES

COPYRIGHT 2015

RADIOACTIVE WASTE THE ALTERNATIVES

‖‖

The Problem:

The nuclear waste is a growing problem that is the byproduct of the nuclear energy. NASA knew that the electromagnetic radiation could be polarized. The example which NASA used to explain this phenomena was the sunglasses which are able to eliminate glare by absorbing the polarized portion of the light. However, NASA did not provide a permanent solution for the Radioactive Waste (electromagnetic radiation).

The Ideas:

The *First observation*: "One of the physical properties of light is that it can be polarized. Polarization is a measurement of the electromagnetic field's alignment. In the figure above, the electric field (in red) is vertically polarized. Think of a throwing a Frisbee at a picket fence. In one orientation it will pass through, in another it will be rejected. This is similar to how sunglasses are able to eliminate glare by absorbing the polarized portion of the light." (http://missionscience.nasa.gov/ems/02_anatomy.html)

The *Second Observation*: "Siegel, (Dennis Siegel, of the University of Arts Bremen), says he has produced two versions of the harvester: One for very low frequencies, such as the 50/60Hz signals from mains power — and another for megahertz (radio, GSM) and gigahertz (Bluetooth/

WiFi) radiation.........The German student has built an electromagnetic harvester that recharges an AA battery by soaking up ambient, environmental radiation. These harvesters can gather free electricity from just about anything, including overhead power lines, coffee machines, refrigerators, or even the emissions from your WiFi router or smartphone. (http://www.extremetech.com/extreme/148247-german-student-creates-electromagnetic-harvester-that-gathers-free-electricity-from-thin-air).

The Third Observation: "Researchers at Georgia Tech scavenged sufficient microwatts to power a temperature sensor, using the ambient energy produced by a television station signal that was a third of a mile away." (http://www.popsci.com/technology/article/2011-07/new-printable-devices-can-harvest-ambient-energy-power-small-electronics)

The Solution:

1. "Currently, there are no permanent disposal facilities in the United States for high-level nuclear waste; therefore commercial high-level waste (spent fuel) is in temporary storage, mainly at nuclear power plants." (http://www.nrc.gov/reading-rm/doc-collections/nuregs/brochures/br0216/).
2. It is important to convert the temporary storage at the nuclear power plants to permanent storage locations.
3. The productions of the electric cars will be the major car productions in the near future and recycling the car batteries and recharging them will be inevitable as a new business.
4. Harvesting the electromagnetic radiation with a coil of wires will be useful for the military and the civil businesses.
5. Storing the electromagnetic radiation materials at the nuclear power plants should be redesigned to include the coil of wires to produce the electrical current.
6. According to this research the Radioactive Waste could be used as an alternative fuel resource for space rockets and Mach 12 Warplanes.

REAL-TIME IMAGERY WEB

COPYRIGHT 2015

REAL-TIME IMAGERY WEB

The Problem

- To improve real-time defense against evolving air and surface combat threats, and to facilitate by extreme precision and an ability to defend against greater numbers of simultaneous and diverse attacks.
- To create a practical way to be in more than one place at a time.
- To reduce the time from calling in a strike to the weapon hitting the target.
- To improve speed and survivability of ground forces engaged with enemy forces.
- To conduct airborne intelligence, surveillance and reconnaissance (ISR) and strike mobile targets anywhere, around the clock.
- To coordinate and direct airstrikes by a ground agent from manned or unmanned air vehicles.
- To enable on-demand, ship-based unmanned aircraft systems (UAS) operations without extensive, time-consuming.

The Solution:

1) At the equator, Earth's circumference is 24,901.55 miles; however, we will need 4 satellites. Each satellite will cover 6225.39 miles from one point to the other until the satellite reach the beginning of the next 6225.39 miles.

2) Each satellite orbiting the earth will start to cover the next distance, 6225.39 miles, which the previous satellite is gradually going through; it would be optimal to have real-time images and video of any place on earth at any time. During the flight in the orbit the four satellites must be in equal intervals (time and distance).

3) Each satellite will be equipped by high tech cameras, videos, radars which will be changing positions gradually from the first mile of the 6225.39 miles to the last mile.

4) The step 3 will be repeated by each of the other 3 satellites.

5) The distance between each two of the four satellites must be equal, assuming the speed of the four satellites are also equal.

6) In this research I assume 4 satellites should be used but the experiment might require more satellites to cover less distances. For example, each one of eight satellites will need to cover 3112.70 miles. That is 24901.55 miles divided by 8 satellites equal 3112.70 miles.

7) The next page will illustrate this solution.

REAL-TIME IMAGERY WEB

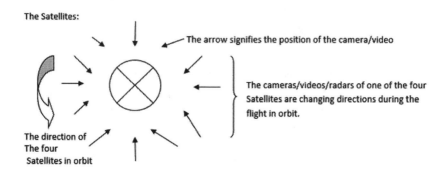

The Satellites:

The arrow signifies the position of the camera/video

The cameras/videos/radars of one of the four Satellites are changing directions during the flight in orbit.

The direction of The four Satellites in orbit

The Satellites:

1) The arrows signify the different positions of the camera/video/ radars from one distance to the next. The radars are used if the sky is cloudy.
2) The circle signifies the earth.
3) Each three arrows signify the continuation of changing directions of the camera/video of one satellite.
4) The speed of the satellites in the orbits and the rotation speed of the earth must be calculated to have the desire results.
5) The real-time images and video of any place on earth at any time will be recorded.
6) The satellite will continue to send the instant images and the videos to the military operation command.
7) The direction of the four satellite in orbit is assumed

REPLACEMENT ANALYSES OF F-35 LIGHTNING II – (EDITED)

COPYRIGHT 2014

REPLACEMENT ANALYSES OF
F-35 LIGHTNING II – (EDITED)

The first problem that is associated with the fighter jet in general is the limited capability of the pilot as a human being and the speed of the fighter jet. The second problem that is associated with the fighter jet is the maximum speed of the fighter jet and the huge distance between the American military bases and the location of the enemy. The third problem that is associated with the fighter jet is the slow response to the satellite signals (pictures) which locate a particular location of the military equipment of the enemy because the slow response is due to the huge distance between the USA military bases and the location of the enemy.

The main concern is to narrow the response time between the signals of the satellite and destroying the enemy. The speed of F-35 is Mach 1.6 but the speed of the rocket is Mach 10 (China is experimenting with Mach 12). When producing the rocket in the same way the F-35 is produced to be able to carry several bombs/guided missiles/clusters, intelligent data, stealth, and satellite networking connections the delivery of the bombs for several enemy locations will be faster and instant. Of course, the rocket will be unmanned so the physical pilot factor will be eliminated. The rocket will be operated remotely via the satellite.

Analyses:

1 Mach [Dry Air @ 273 Kelvin] = 741.81729 miles/hour

If the speed of the rocket is Mack 10 and it is launched from Washington DC, the distance from Washington DC to Damascus = 5877 miles

10 Mach [Dry Air @ 273 Kelvin] = 7,418.1729 miles/hour

5877 / 7,418.1729 = .7922 hour = 60 * .79 = 47 minutes

China is experimenting on Mach 12 rocket, so If the speed of the rocket is Mack 12 and it is launched from Washington DC, the distance from Washington DC to Damascus = 5877 miles.

12 Mach [Dry Air @ 273 Kelvin] = 8,901.80748 miles/hour

5877 / 8,901.80748 = .6602 hour = 60 * .66 = 39.60 minutes

If the speed of the rocket is Mack 10 and it is launched from the United Kingdom - Scotland - Edinburgh, the distance from Scotland – Edinburgh to Damascus = 2433 miles

10 Mach [Dry Air @ 273 Kelvin] = 7,418.1729 miles/hour

2433 / 7,418.1729 = .3279 hour = 60 * .32 = 19.2 minutes

To utilize Mach 12 rocket:

12 Mach [Dry Air @ 273 Kelvin] = 8,901.80748 miles/hour

2433 / 8,901.80741 = .27 hour = 60 * .27 = 16.2 minutes

When the rocket, Mach 12, is launched from the United Kingdom - Scotland – Edinburgh, the rocket will take 16.2 minutes from receiving the satellite (pictures) signals of the enemy's convoy (trucks and tanks) in Damascus to the moment of destroying those trucks and tanks.

In comparison, F-35 that is launched from the United Kingdom - Scotland – Edinburgh will take 2.04 hours to reach Damascus.

1.6 Mach [Dry Air @ 273 Kelvin] = 1,186.907664 miles/hour

2433 / 1,186.907664 = 2.04 hours =60 * 200.04% = 120.24 minutes

REPLACEMENT OF LEGGED SQUAD SUPPORT SYSTEM (LS3)

COPYRIGHT 2014

REPLACEMENT OF LEGGED SQUAD SUPPORT SYSTEM (LS3)

"Today's dismounted warfighter can be saddled with more than 100 pounds of gear, resulting in physical strain, fatigue and degraded performance. Reducing the load on dismounted warfighters has become a major point of emphasis for defense research and development, because the increasing weight of individual equipment has a negative impact on warfighter readiness. The Army has identified physical overburden as one of its top five science and technology challenges. To help alleviate physical weight on troops, DARPA is developing a four-legged robot, the Legged Squad Support System (LS3), to integrate with a squad of Marines or Soldiers" (TTO Programs).

Developing a four-legged robot is a mistake and it is not practical in today's war. I heard today a republican senator is saying that in order for USA to win the war in Syria Mr. President Obama must order ground troops to invade and to face ISIL. This idea of this republican senator shows his out-of-date and primitive ideas about the wars in the 21st century. You do not have to win and dictate your power by using ground troops and robots. Sir, you definitely remember World War II and the two atomic bombs over Japan. Crippling Japan was the key approach to dictate the American demands and it was done by the United States Air Force. The republican senator's ideas are reflecting his sadistic psychology to see the American soldiers killed and injured!!!!!!!! You can, Sir, cripple ISIL using advanced/innovative unconventional weapons not on ground troops and robots.

The first problem that is associated with the fighter jet in general is the limited capability of the pilot as a human being and the speed of the fighter jet. The second problem that is associated with the fighter jet is the maximum speed of the fighter jet and the huge distance between the American military bases and the location of the enemy. The third problem that is associated with the fighter jet is the slow response to the satellite signals (pictures) which locate a particular location of the military equipment of the enemy because the slow response is due to the huge distance between the USA military bases and the location of the enemy.

The main concern is to narrow the response time between the signals of the satellite and destroying the enemy. The speed of F-35 is Mach 1.6 but the speed of the rocket is Mach 10 (China is experimenting with Mach 12). When producing the rocket in the same way the F-35 is produced to be able to carry several bombs/guided missiles/clusters, intelligent data, stealth, and satellite networking connections the delivery of the bombs for several enemy locations will be faster and instant. Of course, the rocket will be unmanned so the physical pilot factor will be eliminated. The rocket will be operated remotely via the satellite.

Analyses:

1 Mach [Dry Air @ 273 Kelvin] = 741.81729 miles/hour

If the speed of the rocket is Mack 10 and it is launched from Washington DC, the distance from Washington DC to Damascus = 5877 miles

10 Mach [Dry Air @ 273 Kelvin] = 7,418.1729 miles/hour

5877 / 7,418.1729 = .7922 hour = 60 * .79 = 47 minutes

China is experimenting on Mach 12 rocket, so If the speed of the rocket is Mack 12 and it is launched from Washington DC, the distance from Washington DC to Damascus = 5877 miles.

12 Mach [Dry Air @ 273 Kelvin] = 8,901.80748 miles/hour

5877 / 8,901.80748 = .6602 hour = 60 * .66 = 39.60 minutes

If the speed of the rocket is Mack 10 and it is launched from the United Kingdom - Scotland - Edinburgh, the distance from Scotland – Edinburgh to Damascus = 2433 miles

10 Mach [Dry Air @ 273 Kelvin] = 7,418.1729 miles/hour

2433 / 7,418.1729 = .3279 hour = 60 * .32 = 19.2 minutes

To utilize Mach 12 rocket:

12 Mach [Dry Air @ 273 Kelvin] = 8,901.80748 miles/hour

2433 / 8,901.80741 = .27 hour = 60 * .27 = 16.2 minutes

When the rocket, Mach 12, is launched from the United Kingdom - Scotland – Edinburgh, the rocket will take 16.2 minutes from receiving the satellite (pictures) signals of the enemy's convoy (trucks and tanks) in Damascus to the moment of destroying those trucks and tanks.

In comparison, F-35 that is launched from the United Kingdom - Scotland – Edinburgh will take 2.04 hours to reach Damascus.

1.6 Mach [Dry Air @ 273 Kelvin] = 1,186.907664 miles/hour

2433 / 1,186.907664 = 2.04 hours =60 * 200.04% = 120.24 minutes

REPLACEMENT OF NUCLEAR WARHEAD MISSILES

COPYRIGHT 2014

This is a wake-up call.

It is known that the Department of Defense is working diligently "to achieve a stealthy and survivable subsonic cruise missile........to engage specific enemy warships from beyond the reach of counter-fire systems" (TTO Programs). I would like to add my voice to this urgent technology and also I might say that the American nuclear warhead missiles are ineffective because the anti-missiles are produced by other countries, probably Russia and China, e.g., and the Russian president Putin is much a reminder of Hitler and the Nazi regime when they began to invade the neighboring countries. Converting the American nuclear warhead missiles to stealth nuclear warhead missiles is the most important/urgent technology we need in our time. The United State of America must be sure also the American nuclear warhead missiles not only stealth but at least 12 times faster than the speed of sound. The speed of the Chinese rockets in China is Mach 12. In the news, Russia already created nine times the speed of sound missiles.

September 20, 2014 and February 2019

RESEARCH IDEAS - NEOLITHIC

COPYRIGHT 2015

"There were no extraterrestrial beings or aliens. The Egyptians definitely built the pyramids and temples. The Lebanese built their million tons columns. The Neolithic, that is, the new stone age was on earth between 10,000-2,000BC and during the ancient Egypt civilization. At this period of time, the earth experienced weak gravity acceleration and force. The Egyptians were able to lift huge and gigantic stones not because of the extraterrestrial machines and equipment but because the gravity acceleration and force were weaker than the gravity acceleration and force that we experience now in this current era." (Milad, 2014).

References

Milad, A. I. (2014). **Neolithic. Copyrighted 2014**

RESEARCH IDEAS

COPYRIGHT 2015

RESEARCH IDEAS

Poisoning Gases

It is true that the plants absorb carbon dioxide and release oxygen during the day. The carbon dioxide and the other harmful gases are converted to the trunks and the leaves of the trees which are solid substances. Accordingly, we are not only causing the pollution in our atmosphere but the earth might be absorbing poisoning gases from the space. We can change or adjust the DNA of some of the plants to absorb more of the poisoning gases and to convert them to solid substances. Probably, we will have substances that are different from the substances of the trees that we all knew, woods. Apparently, woods will be the other kind of the biological matters. But how can we change the DNA that is responsible for absorbing the earthly harmful gases to absorb the gases which we are causing and those are absorbed by the earth from the space? We can change the DNA by creating a liquid switch and to distribute it by airplanes.

Nitrogen

For example, if the nitrogen is the gas that is dominating a distant planet but it is a suitable planet for life, then to travel to this planet and to live on the surface we have to alter our DNA and gene which is responsible for our existence on earth to fit the distant planet.

The Earthquake Project

The earthquake project is brilliant. The project is to explode an underground bomb to shake the earth's crust and to direct the vibrations under the enemy's city. Knowing the earth's crust particular layer, which connect the location of the bomb and the enemy city, the more successful the earthquake project will be.

Electromagnetic Waves

We can increase the magnetic shield around the earth by producing electromagnetic waves from several spots, the north pole and the south pole, on this planet to increase the strength of the natural magnetic shield of the earth to prevent the harmful radiation and the flare of the sun from penetrating the earth atmosphere or destroying our satellites in the earth orbits.

SHUTDOWN MEANS PUNISHING THE AMERICANS

Shutdown Means Punishing the Americans

To shutdown the government is to not pay the employees of the federal government their salaries and wages. The president, any president, could use this policy to achieve his/her political philosophy. For example, building walls at the Mexican border will not stop the illegal immigrants, from the entire world, from entering the USA through the air and the oceans in the east and the west. The jails in the USA are occupied fully by American citizens. To shutdown the government is to punish the honest employees of the federal government by denying them their salaries and wages. This policy does not make sense to me and the American people everywhere. The policy to shutdown the government will fit only in the dictatorship political system. For the sake of humanity, Berlin wall was destroyed and removed as a sign of the democratic philosophy which was adopted by Ronald Reagan. Also, the southern border wall is a reminder of the ancient China wall which made a moderate sense when the wall was built but now with the superior technology it is the symbol of nonsense!!!! To reflect our advancement in this century in the superior technology the government should applies the superior technology to track the illegal immigrants who are coming from the southern border and the four borders in general.

1/25/2019

2/1/2019

SPACE DEBRIS AND THE ARTIFICIAL GRAVITY

SPACE DEBRIS AND THE ARTIFICIAL GRAVITY

The Problem:

- "The number of space debris that threaten important communications, weather monitoring, navigation services and imagery satellites, is growing.
- Debris are threatened more than 1,200 satellites that are currently operated by commercial and government entities around the globe.

Current deep space search telescopes are unable to provide a full picture of objects such as microsatellites and space debris that threaten military satellites" (TTO Programs).

"More than 7,000 spacecraft have been launched from Earth, the vast majority of which are satellites that are no longer operational. These defunct objects, now free-orbiting debris, threaten the more than 1,200 satellites that are currently operated by commercial and government entities around the globe. The number of space debris that threatens important communications, weather monitoring, navigation services and imagery satellites is growing." (TTO Programs).

"The United States Space Surveillance Network (SSN), a worldwide network of 29 space surveillance sensors (radar and optical telescopes) is tasked with observing and cataloging space objects. As new space tracking challenges arise and as a result of lengthy acquisition timelines to bring new sensors online, the network is impeded in its ability to keep pace." (TTO Programs).

Background:

1) For the antigravity, "Someone in the laboratory was smoking a pipe, and the pipe smoke rose in a column above the superconducting disc. So we placed a ball-shaped magnet above the disc, attached to a balance. The balance behaved strangely. We substituted a nonmagnetic material, silicon, and still the balance was very strange. We found that any object above the disc lost some of its weight, and we found that if we rotated the disc, the effect was increased."[1] Platt, Charles (June 3, 1998). "Breaking the Law of Gravity". Wired. Retrieved 2011-06-19. https://en.wikipedia.org/wiki/Eugene_Podkletnov

2) The beam (0f the antigravity)doesn't disappear rapidly with distance – in fact, it's been measured at distances of up to 5 kilometers, and seems to penetrate all materials without a decrease in force. These are only a few of the details that he provides. http://www.pureenergysystems.com/news/2004/08/04/6900035EugenePodkletnov/

3) According to Dr. Eugene Podkletnov, the discovery was accidental. It emerged during routine work on so-called "superconductivity", the ability of some materials to lose their electrical resistance at very low temperatures. The team was carrying out tests on a rapidly spinning disc of superconducting ceramic suspended in the magnetic field of three electric coils, all enclosed in a low-temperature vessel called a cryostat. http://www.scansite.org/scan.php?pid=158

4) In an article in the Sunday Telegraph BREAKTHROUGH AS SCIENTISTS BEAT GRAVITY by Robert Matthews and Ian Sample September 1, 1996, page 3 he is quoted as saying: "One of my friends came in and he was smoking his pipe," Dr. Podkletnov said. "He put some smoke over the cryostat and we saw that the smoke was going to the ceiling all the time. It was amazing—we

couldn't explain it." Tests showed a small drop in the weight of objects placed over the device, as if it were shielding the object from the effects of gravity – an effect deemed impossible by most scientists. "We thought it might be a mistake," Dr. Podkletnov said, "but we have taken every precaution." Yet the bizarre effects persisted. The team found that even the air pressure vertically above the device dropped slightly, with the effect detectable directly above the device on every floor of the laboratory." http://www.scansite.org/scan.php?pid=158

5) For artificial gravity, the conclusion is that you need to keep the spinning environment large compared to the size of a human body (100 meters radius would be nice), to minimize the absolute velocity differences between head and feet, and to keep the number of revolutions per minute relatively low (probably less than 3-4). You also need to configure it so that the actual velocity of the circular motion is significantly larger than the typical speeds that humans are going to walk or run at. Otherwise, if you run in an anti-spin direction you risk negating the very acceleration that's keeping you pushed to the floor!

The bottom line is that, as the Gemini 11 astronauts got a little taste of, not only is setting up an artificial gravity system a bit of an engineering challenge, it's also a challenge to tune things to adequately accommodate what humans (or any other organisms) can cope with. It is not as simple as just spinning up your space wheels. http://blogs.scientificamerican.com/life-unbounded/watch-the-first-artificial-gravity-experiment/

6) For the initial study this summer, 32 test subjects will be placed in a six-degree, head-down, bed-rest position for 21 days to simulate the effects of microgravity on the body. Half that group will spin once a day on the centrifuge to determine how much protection it provides from the bed-rest deconditioning. The "treatment" subjects will be positioned supine in the centrifuge and spun up to a force equal to 2.5 times Earth's gravity at their feet for an hour and then go back to bed.

"The studies may help us to develop appropriate prescriptions for using a centrifuge to protect crews and to understand the side effects of artificial gravity on people," said Dr. Bill Paloski, NASA principal scientist in JSC's Human Adaptation and Countermeasures Office and principal investigator for the project. "In the past, we have only been able to examine bits and pieces. We've looked at how artificial gravity might be used as a countermeasure for, say, cardiovascular changes or balance disorders. This will allow us to look at the effect of artificial gravity as a countermeasure for the entire body," he added. http://www.nasa.gov/home/hqnews/2005/apr/HQ_05109_Artifical_Gravity.html

7) Gravitational acceleration was constant both times you dropped the items. The only difference was the presence of air mass acting upon the feather: because the object is of low density, the feather encounters more drag as it falls through the air. This experiment shows us that weight doesn't determine the rate at which something falls—only air resistance does.

http://www.education.com/science-fair/article/feather-coin/

8) In a famous demonstration, Galileo supposedly dropped a heavy weight and a light weight from the top of the Leaning Tower of Pisa to show that both weights fall at the same acceleration. However, this rule is true only if there is no air resistance. This demonstration lets you repeat Galileo's experiment in a vacuum. http://www.exploratorium.edu/snacks/falling-feather

The Idea:

1) Connecting the two experiments (above) will lead us to say that the debris will be in vacuum where no air resistance and spinning the inner cylinder (that is suggested by this project)

will create an artificial gravity that will gravitate all types of debris that are floating in the earth orbit in space.

The Solution:

1) The solution is to convert an old shuttle, or creating a new space shuttle, to be the new Space Debris Collector (SDC). Besides observing and cataloging space objects, the space debris needs to be removed from the orbit of this planet. Removing the space debris must become a routine work as long as we are using the orbit for satellites. In the future, the satellites should be equipped mechanically to return to the ground space station and reused.

2) The Space Debris Collector (SDC) should be equipped with a large cylinder inside a larger cylinder.

3) The inner cylinder will be designed to rotate as a centrifuge to collect the debris in space.

4) Creating antigravity by Dr. Eugene Podkletnov opens the possibility for Artificial Gravity.

5) Galileo supposedly dropped a heavy weight and a light weight from the top of the Leaning Tower of Pisa to show that both weights fall at the same acceleration. However, this rule is true only if there is no air resistance. This demonstration lets you repeat Galileo's experiment in a vacuum.

6) Rotating the inner cylinder as I mentioned in # 2 will be effective for all types of debris and all materials in the earth orbit because the experiment will be conducted in space and in vacuum where no air resistance.

7) The inner cylinder which is full with debris will be released in the atmosphere of the Sahara Desert in North Africa.

8) the Shuttle will return to NASA's base to replace the inner cylinder.

9) NASA is determined to use the artificial gravity on people especially the astronauts who will be traveling in space but this project will be manageable and easier for the space debris that are caused by our technology in the earth orbit.

SPACE DEBRIS COLLECTOR

COPYRIGHT 2013

SPACE DEBRIS COLLECTOR

The Problem:

"More than 7,000 spacecraft have been launched from Earth, the vast majority of which are satellites that are no longer operational. These defunct objects, now free-orbiting debris, threaten the more than 1,200 satellites that are currently operated by commercial and government entities around the globe. The number of space debris that threatens important communications, weather monitoring, navigation services and imagery satellites is growing." (TTO Programs).

"The United States Space Surveillance Network (SSN), a worldwide network of 29 space surveillance sensors (radar and optical telescopes) is tasked with observing and cataloging space objects. As new space tracking challenges arise and as a result of lengthy acquisition timelines to bring new sensors online, the network is impeded in its ability to keep pace." (TTO Programs).

The Solution:

The solution is to convert an old shuttle, or creating a new space shuttle, to be the new Space Debris Collector (SDC). Besides observing and cataloging space objects, the space debris needs to be removed from the orbit of this planet. Removing the space debris must become a routine work as long as we are using the orbit for satellites. In the future, the satellites should be equipped mechanically to return to the ground space station and reused.

The Idea:

1) The maneuver of the space shuttle in the orbit needs to be improved.
2) The astronaut of the space shuttle needs to be able to navigate the SDC in all directions.
3) The space shuttle will be equipped with robotic arms.
4) The robotic arms will grab the debris, one object at a time, and bring it inside the shuttle.
5) The shuttle will smash the debris and compress it to make room for other debris.
6) When the container of the shuttle is packed with the compressed debris the shuttle will descend to the ground space station.
7) It is our responsibility to collect the defunct objects to keep the orbit safe.
8) The SDC will return to the orbit again and again to collect the space debris
9) Imagine the Space Debris Collector (SDC) is the Recycling Truck of the 21st century.
10) The debris that is collected from the orbit could be reused or recycled.
11) Enclosed are two patents which might be helpful to creating steering control for SDC or they might be helpful to creating the robotic arms.

SPACE DEBRIS POTENTIAL REMOVER

COPYRIGHT 2015

SPACE DEBRIS POTENTIAL REMOVER

The Problem:

- "The number of space debris that threaten important communications, weather monitoring, navigation services and imagery satellites, is growing.
- Debris are threatened more than 1,200 satellites that are currently operated by commercial and government entities around the globe.
- Current deep space search telescopes are unable to provide a full picture of objects such as microsatellites and space debris that threaten military satellites" (TTO Programs).

The Solution:

1) The solution must be a continuous operation as long as the satellites are launched and orbiting the Earth.
2) Using a mile of a "pizza box" as suggested by NASA is not practical.
3) Using a magnetic field to collect the debris is not practical and useless.
4) The solution this research suggesting is to force the debris to be gravitated toward the earth.
5) Launching a spacecraft that is designed to include an 8 ft by 8 ft "cylinder" will be the probable measure for this research.
6) The suggested spacecraft will only operate in orbit over the deserts around the globe.
7) Inside the cylinder an explosion will be produced.

8) Hydrogen gas or gasoline will be released in the cylinder and a spark will ignite the flammable gases to push the debris toward the desert.

9) The spark will take a second and the explosion force will be enough to force the debris toward the desert and not enough to move the spacecraft the opposite direction.

10) Repeating step (8) continuously during the travel over the desert will force the debris to be gravitated toward the earth/ desert.

11) This particular spacecraft must be equipped to be reused or rendezvous to be refilled with the flammable gases.

12) The suggested spacecraft will be equipped by a telescope, camera, and video to provide a full picture of objects which are removed or remained such as microsatellites and space debris that threaten military satellites.

SPIDERS FOR MASSIVE SCALES

COPYRIGHT 2015

SPIDERS FOR MASSIVE SCALES

The Problem:

"Current defense systems for processing information struggle to effectively scale to the volume and characteristics of changing data environments and the range of applications for data analysis. Overcoming these challenges requires fundamentally new approaches to data science, including distributed computation and interactive visualization.......for processing and analyzing large, imperfect and incomplete data." (Wade, Information Innovation Office (I2O)) (DARPA-BAA-14-39)

Outline:

Spiders are referred to ships and drones. The ships will be located nearby the terrors' spots around the world and places for the drones to be launched and returned. Ships will be a floating basis for the drones.

The Solution:

1) The ships will be nearby the terror's locations and the hot spots in the globe to listen to the telephones and the cellular phones through the satellites.
2) The drones will be launched and directed to the source of the suspicious calls.
3) For example, ISIS is located in El Yemen; however, the ship which is specialized to listen to the Arabic accent of the people of El Yemen will be the informer of the suspicious conversation

and the location will be determined by the cellular phones. At this moment the drones will be sent to hover over this location.

4) The drones will be equipped by high tech cameras, weapons, radar, and communication capabilities.

5) The spiders, ships and drones, will be equipped with the Navy protection ability.

6) Each spider will be specialized in an accent of the enemy's language to be able to understand and to translate the language to English.

7) The software on the spider ship will be capable of translating, filtering, and locating the conversations in the suspected locations through the phones conversations.

8) The key word is that the ship must focus the intelligence to the enemy's location.

9) By narrowing the location of the enemy to one location the result will lead to fewer telephone conversations and the listening will be manageable.

10) The spider drones which are released over the enemy location in high altitude will circulate the enemy geographic location for 36 hours to take real-time images and videos.

11) The military drones should be used. If we know the military drone last 36 hours flying.

12) The enemy suspects/troops will not be able to remain in the same hiding place for 36 hours.

13) If the enemy suspects/troops were able to hide for 36 hours the first wave of drones will be replaced by other wave of drones that will continue the real-time images and videos.

14) It is like the saying "the chicken or the eggs came first." The spider ship or the spider drone could be first to locate the enemy and take real-time images.

15) When the first wave of drones are about to leave the enemy geographic location after about 36 hours the second wave will join the first wave to make sure our military did not miss the enemy movement and relocation.

STRAIGHT GAY AND LESBIAN
A MILITARY ANALYSIS

COPYRIGHT 2016

STRAIGHT GAY AND LESBIAN
A MILITARY ANALYSIS

The Problem:

The human creation/formation is a contradiction and inconsistency story. The Bible says : "God created man in the image of himself, in the image of God he created him male and female he created them." (The Jerusalem Bible, Reader's Edition, 1968).

The Idea:

Creating male and female in God's image rises a question, which is, How did God create female and male at the same time in his image? How is it possible?

The Facts:

1) The male has estrogen and testosterone.
2) The female has estrogen and testosterone.
3) The normal male has a sexual fantasy to have sex with another male and female.
4) The normal female has a sexual fantasy to have sex with another female and male.
5) The average male does not understand the straight male is also has a sexual fantasy to have sex with another male.
6) The average female does not understand the straight female is also has a sexual fantasy to have sex with another female.

7) The so called straight is not a gay or a lesbian until s/he allows himself or herself to act sexually with the same sex.

8) The average gay and lesbian do not understand that their sexual fantasy is identical to the fantasy of the straight man or straight woman.

9) The average gay or lesbian does not know their fantasy is identical to the fantasy of the straight male and female.

10) The abnormality feeling the gay and lesbian suffer because they think they are not normal.

11) The story of the Bible about Sodom and Gomorrah should not happen. Also David should not kill another man to have sex with the victim's wife. In addition, Moses should not murder the Egyptian. These stories are unethical and it is recorded in our own Bible.

12) When God created Adam and Eve they both were in God's image. The sexual hormones in both Adam and Eve are identical and there was no reason to destroy Sodom and Gomorrah.

13) Mass murderer Omar Mateen in Orlando -Florida seemed to struggle between his female and male fantasies. He Killed gays and lesbians to tell the world he is a man and does not have the female sexual fantasy!!

14) It is clear that the contradiction and inconsistency between the religion teaching and his reality as human who carry the male and female hormones was severe in his mind.

STRATEGIC TACTICS FOR ALTERNATIVE WEAPONS

410-282-8708
COPYRIGHT 2015

STRATEGIC TACTICS FOR ALTERNATIVE WEAPONS

The Problem:

1) There is a difficulty to locate the enemy's scattered troops in the city, the mountain, or the forest.
2) The conventional weapon will not detect/locate human beings in caves, fox holes, or in buildings.
3) The means of communications such as wireless telephones, iPads, and computers are major elements for information in the war zone in the city, the mountain, or the forest.
4) The conventional weapons will not be able to destroy every means of communications instantly in the war zone in the city, the mountain, or the forest.

The Solution:

The Strategies are:

1) "We are all familiar with the power of electromagnetic attacks from the movies: in Ocean's Eleven, George Clooney's gang disables Las Vegas' power grid, and Keanu Reeves' henchmen hold off the enemy robot fighters from their spaceship in the Matrix Trilogy." https://www.fraunhofer.de/en/press/research-news/2013/december/Defending-against-electromagnetic-attacks.html

2) The strategy is that the equipment which produces the high frequency electromagnetic signals should be installed in a non-electrical/computer war plane.

3) The high frequency electromagnetic signals that are aimed toward the war zone will cover several miles and it will penetrate walls and caves.

4) The high frequency electromagnetic signals will cause moderate harm to the enemy troops but will destroy the enemy electronic equipment permanently.

5) Creating non-electrical/computer war planes that are equipped/specialized to produce high frequency electromagnetic signals could be the solution to cripple/confuse the enemy troops when permanently creating electrical interference in the enemy's electronic offense/defense systems.

6) To know precisely the location of each individual terrorist will not be necessarily because the high frequency electromagnetic signals will be effective for miles.

7) The purpose of this project concept and idea is to cripple the enemy's communication equipment and to create/generate confusions within the enemy troops.

SUBSTITUTE OF TITANIUM

COPYRIGHT 2015

SUBSTITUTE OF TITANIUM

The Problem:

The first problem is that "The United States did not have enough Titanium to build the fleet Blackbird and ironically, we bought the needed Titanium from Russia." (http://www.wvi.com/~sr71webmaster/ srqt~1.htm). Our dependence on Russia to build the airframe creates severe setback.

The second problem is the combination of the atmosphere and the high speed of the warplanes. Using the Titanium to build the airframe will give the capability to the warplane to fly in the atmosphere without the meltdown of the airframe. Because the Titanium is a Russian product we have to have other alternatives.

The Idea for Mach 12 Warplane:

The main concern is to narrow the response time between the signals of the satellite and destroying the enemy and to produce the airframe with a high melting point, 4,730 degrees Fahrenheit. The speed of Blackbird is Mach 3 but the speed of the rocket is Mach 10 (China is experimenting with Mach 12). When producing the rocket as Mach 12 Warplane in the same way the Blackbird is produced to be able to carry several bombs/guided missiles/clusters, intelligent data, stealth, and satellite networking connections the delivery of the bombs for several enemy locations will be faster and instant. As an alternative, the suggested Mach 12 Warplane will be unmanned so the physical pilot factor will be eliminated and it will be operated remotely via the satellite. As another alternative, the Mach 12 Warplane will be operated by a pilot who will be wearing a space suit and inside a rocket designed cockpit. The rocket fuel pumps should be adjustable to control the speed and maneuvers.

The Idea for Blackbird:

The substitute of Titanium is Molybdenum which "the U.S. produces significant quantities of molybdenite. Molybdenum (element #42, symbol Mo) is a metallic, lead-gray element, with a high melting point, 4,730 degrees Fahrenheit. This is 2,000 degrees higher than the melting point of steel, and 1,000 degrees higher than the melting temperature of most rocks. Molybdenum was discovered by Peter Hjelm in 1781" (https://www.mineralseducationcoalition.org/minerals/molybdenum)

Basic Information: (http://www.chemicalelements.com/elements/mo.html)
Name: Molybdenum
Symbol: Mo
Atomic Number: 42
Atomic Mass: 95.94 amu
Melting Point: 2617.0 °C (2890.15 K, 4742.6 °F)
Boiling Point: 4612.0 °C (4885.15 K, 8333.6 °F)
Number of Protons/Electrons: 42
Number of Neutrons: 54
Classification: Transition Metal
Crystal Structure: Cubic
Density @ 293 K: 10.22 g/cm3
Color: silverfish

Molybdenum Statistics and Information:

Molybdenum (Mo) is a refractory metallic element used principally as an alloying agent in steel, cast iron, and superalloys to enhance hardenability, strength, toughness, and wear and corrosion resistance. To achieve desired metallurgical properties, molybdenum, primarily in the form of molybdic oxide or ferromolybdenum, is frequently used in combination with or added to chromium, manganese, niobium, nickel, tungsten, or other alloy metals. The versatility of molybdenum in enhancing a variety of alloy properties has ensured it a significant role in contemporary industrial technology, which increasingly requires materials that are serviceable under high stress, expanded temperature ranges, and highly corrosive environments. Moreover, molybdenum finds significant usage as a refractory metal in numerous chemical applications, including catalysts, lubricants, and pigments. Few of molybdenum's uses have acceptable substitutes. (http://minerals.usgs.gov/minerals/pubs/commodity/molybdenum/)

The solution for Blackbird:

The United States have enough Molybdenum to build the fleet Blackbird and ironically we do not need Titanium from Russia.

The Solution for Mach 12 Warplane:

1. Mach 12 Warplane must be launched to the space orbit to reduce the time traveling in the atmosphere and to avoid the airframe meltdown. "The compression of the air layers near the leading edges of the shuttle is quick, causing the temperature of the air to rise to as high as 3000 degrees Fahrenheit!"(https://www.uu.edu/dept/physics/scienceguys/2003Mar.cfm). But by using Molybdenum with a high melting point, 4,730 degrees Fahrenheit Mach 12 Warplane and the Blackbird will overcome the atmosphere airframe meltdown.

2. If we know the reentry temperature is 3000 degrees Fahrenheit then we can choose steel/ Molybdenum combination for the airframe of the warplane or the Blackbird. The melting point for the steel/ Molybdenum combination is 4742.6 °F. No adjustable speed should be considered to slow down the warplane to reduce the compression of the air near the leading edges of the warplane.

3. Mach 12 Warplane will be launched from any military base around the globe regardless of the bad weather here or in other part of the world because Mach 12 Warplane will be launched to the earth orbit in a few minutes in a clear sky or not as a rocket.
4. Mach 12 Warplane will be operated without pilot or by a pilot(s) who will drive the Mach 12 Warplane to ascend to orbit and to descend over the enemy territory.
5. Mach 12 Warplane will be fueled for the distance from the moment the rocket is ignited until returning safely to the base.
6. Mach 12 Warplane will start by flying as a high speed rocket.
7. Mach 12 Warplane will be reused indefinitely as long as no maintenance is required.
8. The warplane rocket will be refueled with liquid fuel or other suitable fuel.
9. The rocket fuel pumps should be adjustable to control speed and maneuvers.
10. In wars we must break the international laws and treaties. Using the space to win the war is logical.

THE CONSTITUTION ILLITERACY OF RELIGION

COPYRIGHT 2016

THE CONSTITUTION ILLITERACY OF RELIGION

The Problem:

The religions are playing an important role in wars, crisis, chaos, hate, and discriminations. The freedom of religions in the United States should be examined.

The Idea:

The teaching of the religion by the religion clerics to encourage the hateful verses in the temples, churches and mosques must first stop before applying the concept of the freedom of religions. As long as the hateful verses are taught in the worship places the fanatics and extremists will continue to rise.

The Facts:

1) Judaism claims that Jewish people are the chosen people constantly in the temples.
2) Islam claims that Muslim people are the chosen people constantly in the mosques.
3) Judaism claims that King David in the Old Testament is a man of God while he is a murderer who killed a husband to have sex with his wife.
4) Islam claims that Jesus is not the Son of God and he is an ordinary prophet.
5) Islam claims that God is one who has no son.

6) Christianity claims that by Jesus the universe was created and he is the alpha and the omega.
7) Islam claims that Christianity is a false religion because Christians are "moshrikeen" that is the Christians are polytheists.
8) Judaism and Islam claim the woman who committed adultery should be stoned to death.
9) Christianity claims the woman who committed adultery should be forgiven.
10) Repeating these ideas (and other negative ideas are not mentioned in this research) constantly in the temples, churches, and mosques create radical people and fanatics who are willing to die for God's teaching.
11) Educating the clerics in each religion to avoid repeating the discriminatory verses must be done and those discriminatory verses must be included in our legal system as illegal verses and against our national security.
12) The discriminatory/hateful verses must be considered illegal and must not be taught in the temples, churches and mosques or any public places to fuel people's emotions against the other religions.

THE FUTURE STATE OF MEXICO

COPYRIGHT 2014

THE FUTURE STATE OF MEXICO

The USA military must take a strong position toward Mexico. The natural progress between the United States and Mexico has just begun. Please open the border between the United States and Mexico for the natural process/progress to take place during this generation. Mexico is the future State of Mexico that will be included under the federal government of the United States of America. Opening the border between Mexico and the USA will facilitate the union between the two countries. Human beings have the right to emigrate from one geographic area to another. The open border between the two countries means no passport needed for Mexicans to enter the USA and no passport needed for the US citizens to enter Mexico. Normalizing the border and granting the US citizenship to all Mexicans will be the first steps toward the union between the two countries. Offering the economic support to Mexico, the protections of the US federal government and the military might, and relocating the US citizen to Mexico, and nationalizing the Mexicans in the USA will create the humanitarian environment. Mexico cannot avoid the natural progress between the two counties. The natural resources in Mexico with the help of the technology that is developed by the USA will combine the natural resources with the labor force that will form a new powerful economic trend. Building modern trains between Mexico and the USA will facilitate the union between the two countries. Please tear down the walls that are in our borders dividing Mexico and the United States. The natural process will allow the Mexican people to emigrate gradually to the United States which will negotiate the expected union of Mexico and the United States. The Mexican labor and the natural resources of Mexico must be taken seriously. Mexico is now ready to be a part of the USA as the State of Mexico.

6/21/2014

Reference: Milad, A. I. (2014). **Letters to the President as of September 3, 2014.**

THE SCIENTISTS UNDER THE MICROSCOPE

COPYRIGHT 2015

Dear **DARPA-BAA-14-42**

The unscientific science is in the media 24/7 and the scientists are repeating the incorrect information constantly. Here are some examples:

The sun's surface is developing a black hole.

There are millions of the earth-like planets.

Mars is our future planet.

We are looking for a planet which can sustain life.

From my observations here are the true outlooks:

The reality is that the black spot on the surface of the sun is the formation of heavy/solid elements matters which will be discharged in an orbit around the sun and will form a new planet.

There is no earth-like planet in the universe because the earth is the unique creation of our sun and it was discharged from our sun. The human is the evolution of one-cell which is produced in the core of the earth and found in the volcanoes' lava that is more than 2000 Fahrenheit.

It is a well known scientific fact that the universe is expanding. At any moment, hours or a thousand year, the earth will be moved farther from the sun and approximately to Mars orbit and the domino effect in the Solar System will take its course from Mercury to Neptune. The Earth will be frozen in the new distance and Venus will be cooling and it will be the possible planet for life. Mars will be moving farther in the Solar System and its environment will not be suitable for life. It is a matter of time. We need to study Venus and to monitor this planet because it is our, as a human species, only hope to survive in this universe.

We will need to change our DNA to fit the other planet's environment. Looking for a planet that is identical to the earth is not possible. The earth is created by our sun. When the sun nuclear interactions produce heavy/solid elements the sun will discharge these elements in the orbit around the sun. These heavy/solid elements and other gases will generate/construct a new planet. The earth is the sun's creation.

Thank you

TITANIUM SETBACK AND THE ALTERNATIVES

410-282-8708
COPYRIGHT 2015

TITANIUM SETBACK AND
THE ALTERNATIVES

The Problem:

The first problem is that "The United States did not have enough Titanium to build the fleet Blackbird and ironically, we bought the needed Titanium from Russia." (http://www.wvi.com/~sr71webmaster/srqt~1.htm). Our dependence on Russia to build the airframe creates severe setback.

The second problem is the combination of the atmosphere and the high speed of the warplane. Using the Titanium to build the airframe will give the capability to the warplane to fly in the atmosphere without the meltdown of the airframe. Because the Titanium is a Russian product we have to have other alternatives.

The Idea:

1) The main concern is to narrow the response time between the signals of the satellite and destroying the enemy. The speed of Blackbird is Mach 3 but the speed of the rocket is Mach 10 (China is experimenting with Mach 12). When producing the rocket as Mach 12 Warplane in the same way the Blackbird is produced to be able to carry several bombs/guided missiles/clusters, intelligent data, stealth, and satellite networking connections the delivery of the bombs for several enemy locations will be faster and instant. As an alternative, the

suggested Mach 12 Warplane will be unmanned so the physical pilot factor will be eliminated and it will be operated remotely via the satellite. As another alternative, the Mach 12 Warplane will be operated by a pilot who will be wearing a space suit and inside a rocket designed cockpit.

2) The substitute of Titanium is Molybdenum which "the U.S. produces significant quantities of molybdenite. Molybdenum (element #42, symbol Mo) is a metallic, lead-gray element, with a high melting point, 4,730 degrees Fahrenheit. This is 2,000 degrees higher than the melting point of steel, and 1,000 degrees higher than the melting temperature of most rocks. Molybdenum was discovered by Peter Hjelm in 1781" (https://www.mineralseducationcoalition.org/minerals/molybdenum)

The solution: '

1. Mach 12 Warplane must be launched to the space orbit to reduce the time traveling in the atmosphere and to avoid the airframe meltdown. "The compression of the air layers near the leading edges of the shuttle is quick, causing the temperature of the air to rise to as high as 3000 degrees Fahrenheit!"(https://www.uu.edu/dept/physics/scienceguys/2003Mar.cfm).

2. If we know the reentry temperature is 3000 degrees Fahrenheit then we can choose steel/carbon combination for the airframe of the warplane. The melting point for the steel/carbon combination is 2800 degrees Fahrenheit. Adjustable speed should be considered to slow down the warplane to reduce the compression of the air near the leading edges of the warplane. Or we can choose Molybdenum (element #42, symbol Mo) which is a metallic, lead-gray element, with a high melting point, 4,730 degrees Fahrenheit

3. Mach 12 Warplane will be launched from any military base around the globe regardless of the bad weather here or in other part of the world because Mach 12 Warplane will be launched to the earth orbit in a few minutes in a clear sky or not as a rocket.

4. Mach 12 Warplane will be operated without pilot or by a pilot(s) who will drive the Mach 12 Warplane to ascend to orbit and to descend over the enemy territory.
5. Mach 12 Warplane will be fueled for the distance from the moment the rocket is ignited until returning safely to the base.
6. Mach 12 Warplane will start by flying as a high speed rocket.
7. Mach 12 Warplane will be reused indefinitely as long as no maintenance is required.
8. The warplane rocket will be refueled with solid fuel or other suitable fuel.
9. In wars we must break the international laws and treaties. Using the space to win the war is logical.

TRANSIENT CONVERTIBLE WARPLANE

COPYRIGHT 2015

TRANSIENT CONVERTIBLE WARPLANE

The Problem:

1. The current warplanes are unable to maneuver in 90 degrees and they must fly miles to make U turn.
2. The inability to make 90 degrees U turn will force the warplane to consume more fuel.
3. The inability to make 90 degrees U turn will force the pilot to expose his life and the warplane to the incoming enemy missiles.
4. The current jet warplanes are unable to land/lift vertically which should be another option/alternative for the pilot.

The solution:

1. **The Transient Convertible Warplane** will allow the warplane the conversion from jet forward warplane to vertical warplane.
2. For **The Transient Convertible Warplane** to be able to make 90 degrees and U turn the horizontal propellers of both wings must be vertical and tilted in the same direction to rotate the warplane to the right or to the left.
3. The ability to make 90 degrees and U turn will force the warplane to consume less fuel.
4. The ability to make 90 degrees and U turn will force the pilot to save his life and the warplane from the incoming enemy missiles.
5. **The Transient Convertible Warplane** will be able to land/lift vertically which should be another option/alternative for the pilot if necessarily.

UNCONVENTIONAL WAR TACTICS

COPYRIGHT 2013

UNCONVENTIONAL WAR TACTICS

Dear Mr. President Obama   1/5/2013

NASA should be the future defense organization. It is a mistake to discontinue the production of the space shuttles. Instead of discontinuing the production of the space shuttles, the space shuttle should be improved and armed with atomic warheads. The distance from the ground to the orbit around the earth will approximately take 2 minutes. The shuttle in the orbits will reach China in approximately 3 minutes. If we are attacked by China we will need to respond swiftly and instantly and the armed space shuttle will accomplish this goal.

Dear Mr. President Obama   1/13/2013

There are limitations which are occurred by the gravity and we can apply these rules to the Air Force planes. The technology will not reach the ultimate speed without jeopardizing the life of the pilots. The more speed the airplane will reach the human bodies will be shattered. The problem is that in the war situations the Air Force airplane will travel from point A to point B for several hours to attack the enemies. The airplanes whether Air Force planes, Air Force One, or domestic and passenger airplanes are an old technology and obsolete. Your administration, precisely NASA, discontinued the productions of the space shuttles. Instead of improving the space shuttles to be launched without the rocket boosters, NASA discontinued the space shuttles. The space shuttles should be produced as Air Force planes, Air Force One, and domestic and passenger airplanes. The space shuttles are important to travel from point A to point B and from one nation to another nation in a short time. The travel does not have to be from one planet to another planet in our galaxy or in the universe but the shuttle will help to reduce the time to travel to minutes and seconds. Instead of discontinuing the productions of the space shuttles, the space shuttles should be improved. The distance from the ground to the orbit around the earth will approximately take 2 minutes. The shuttle in the orbits will reach China in approximately 3 minutes. If we are attacked by China or Russia we will need to respond swiftly and instantly and the armed space shuttle will accomplish this goal. The capacity of current airplanes is limited and obsolete. A side note, please do not reduce the nuclear warheads. The reason for creating monster destruction bombs is not clear and there is probability that human collectively will need these warheads in the future.

USA WAR TACTICS

COPYRIGHT 2014 - 2015

Because I am an American citizen, I think the war tactics in the 21st century should be changed. If the president of China is giving a speech to declare war against the USA in the Tiananmen Square, the new USA war tactics should bomb the Tiananmen Square and the Chinese president before the Chinese president finishes his sentence. Here is how: NASA should be a future defense organization. It is a mistake to discontinue the production of the space shuttles. Instead of discontinuing the production of the space shuttles, the space shuttle should be improved and armed with atomic warheads. The distance from the ground to the orbit around the earth will approximately take 2 minutes. The shuttle in the orbits will reach China in approximately 3 minutes. If Chinese president declares war against the USA from Tiananmen Square we will need to respond swiftly and instantly and the armed space shuttle will accomplish this goal before the Chinese president finishes his sentence.

Experimentation of a Steering Control for a Spaceship

By

Dr. Anis I. Milad

Copyright 2013

Brief Abstract

This research is an experiment of a patent, Articulatable Mechanism, to investigate the capability of the device in space. The device is intended to perform a U turn in space, 360 degrees, rotation, and tilting in any direction. The patent was filed on August 20, 1990 and it was issued on June 9, 1992. The steering control is a device that is designed to be used in a spaceship, rocket, robot, torpedo, submarine, or aircraft jet or propeller. It is designed to be light weight and easily be steered. The device could be used manually, programmed, or remotely operated. It will help create new technology in the aerospace industries and elsewhere.

Statement of Problem

The recent space shuttle, Discovery, e.g., is designed to have wings assuming the shuttle will be used partly in the Earth's atmosphere. The problem is that the wings increase the risk of crashing and they are not useful during the travel in space. The landing device may not take the shape of wings. The steering control that is experimented in this research might totally terminate the wings; instead, the steering control, which is suggested in this research, will allow the spaceship to land without the need for wings automatically or manually.

Research Question

Can the suggested steering control be used in the spaceship, rocket, robot, torpedo, and submarine to accelerate its maneuverability and to add a U turn capability to the vehicle without the use of wings?

Background and Relevance to Previous Work

In A Review of Jet Mixing Enhancement for Aircraft Propulsion Applications by K Knowles and A J Saddington there is a description of "Jet oscillation which was achieved by using a rotary insert near the nozzle exit. This was operated by a crank mechanism and acted to oscillate the jet exit angle, thus flapping the jet bodily."

In Robust Gain-scheduled Control for Vertical/Short Take-off and Landing Aircraft in Hovering with Time-varying Mass and Moment of Inertia By S-LWu, P-C Chen, K-Y Chang, and C-C Huang a description of "The nozzles are capable of rotating together from the aft position forward approximately 100°. This change in the direction of thrust allows the aircraft to operate in two modes of wing-borne forward flight and jet-borne hovering as well as to transit between them."

In Space Mission Planning and Operations by V. Adimurthy, M. Y. S. Prasad, and S. K. Shivakumar stated that "For a rocket to achieve the mission, it is often necessary continuously to change the thrust orientation. This is known as the steering program."

General Methodology and Procedure and

Explanation of New and Unusual Techniques

A prototype was built by the researcher. The mechanism is applicable to wide variety of situations such as a spaceship, rocket, robot, torpedo, and submarine to accelerate its maneuverability and to add a U turn capability to the vehicle without the use of wings. The device provides a rotational component as well as pivotal component providing practical propulsion in polar coordinates instead of customary linear transmission.

Expected Results and Their Significance and Application

The successful experiment on this device will lead to change in the steering control for the spaceship, rocket, robot, torpedo, and submarine and will add a U turn capability to the vehicle without the use of wings.

References

Adimurthy, V., Prasad, M., and Shivakumar, S. (2007). *Space Mission Planning and Operations*. Indian Academy of Sciences

Knowles, K. & Saddington, A. (2005). *A review of jet mixing enhancement for aircraft propulsion applications* Journal of Aerospace Engineering.

Milad, A. (1992). Articulatable Michanism. United State Patent

Wu, S-L., Chen, P-C., Chang, K-Y., & Huang, C-C. (2008). *Robust Gain-scheduled Control for Vertical/Short Take-off and Landing Aircraft in Hovering with Time-varying Mass and Moment of Inerti.* Journal of Aerospace Engineering.

United States Patent [19]

Milad

[11] **Patent Number:** **5,119,753**

[45] **Date of Patent:** **Jun. 9, 1992**

US005119753A

[54] **ARTICULATABLE MECHANISM**

[76] Inventor: **Anis I. Milad,** 2938 Yorkway, Baltimore, Md. 21222

[21] Appl. No.: **569,407**

[22] Filed: **Aug. 20, 1990**

[51] Int. Cl.⁵ **B63G 8/16**

[52] U.S. Cl. .... **114/338; 74/665 B; 901/29; 440/63**

[58] **Field of Search** .... 901/18, 25, 29; 74/661, 74/665 B, 665 L, 665 G, 665 N; 244/51, 52, 66, 76 J; 475/1, 2; 239/265.35, 587; 114/338, 337; 440/58–63, 53

[56] **References Cited**

U.S. PATENT DOCUMENTS

1,409,850	3/1922	Haney	244/51 X
4,501,522	2/1985	Causer et al.	901/25 X
4,732,106	3/1988	Milad	440/63 X
4,907,937	3/1990	Milenkovic	901/29 X

Primary Examiner—Edwin L. Swinehart
Attorney, Agent, or Firm—Leonard Bloom

[57] **ABSTRACT**

This invention is an articulatable mechanism applicable to a wide variety of situations. The device provides a rotational component as well as a pivotal component providing practical propulsion in polar coordinates instead of customary linear transmission. This device could be used for steering propulsion devices by controlling the orientation and direction of a propeller shaft or rocket exhaust. Another application of this device might be orienting and directing devices such as fire hoses and lasers.

12 Claims, 9 Drawing Sheets

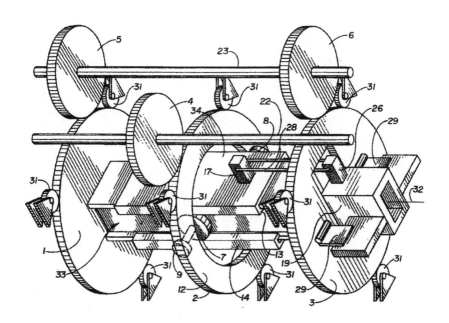

United States Patent [19]

Milad

[11] **Patent Number:** **4,732,106**

[45] **Date of Patent:** **Mar. 22, 1988**

[54] **STEERING CONTROL FOR SUBMARINES AND THE LIKE**

[76] Inventor: **Anis I. Milad,** 2938 Yorkway, Baltimore, Md. 21222

[21] Appl. No.: **887,921**

[22] Filed: **Jul. 22, 1986**

[51] Int. Cl.⁴ B63G 8/16; B63H 5/12
[52] U.S. Cl. 114/338; 74/89.15;
74/665 B; 219/121 LU; 239/265.35; 239/587;
244/51; 244/52; 440/58; 440/63; 901/25
[58] Field of Search 114/337, 338;
440/58–60, 63; 244/51, 52, 66, 76 J, 169, 74;
239/265.35, 587; 74/665 B, 665 A, 89.15; 121
901/25; 219/121 LU, 121 LV, 121 LW, 121
LX

[56] **References Cited**

U.S. PATENT DOCUMENTS

1,409,850	3/1922	Haney	244/51 X
2,131,155	9/1938	Waller	244/51
3,809,318	5/1974	Yamamoto	239/587 X
4,281,795	8/1981	Schweikl	239/265.35
4,497,319	2/1985	Sebine et al.	219/121 LU
4,501,522	2/1985	Causer et al.	901/25 X
4,542,278	9/1985	Taylor	219/121 LV
4,578,554	3/1986	Coulter	219/121 LV
4,579,299	4/1986	Lavery et al.	239/265.35 X

Primary Examiner—Sherman D. Basinger
Attorney, Agent, or Firm—Leonard Bloom

[57] **ABSTRACT**

A steering apparatus for steering propulsion devices such as a propeller shaft and rocket exhaust and for aiming devices such a fire hoses and lasers. The steering apparatus provides a rotational component and a tilting component to the orientation of the steered device.

13 Claims, 16 Drawing Figures

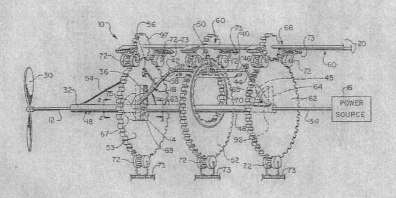

Certificate of Registration

This Certificate issued under the seal of the Copyright Office in accordance with title 17, *United States Code*, attests that registration has been made for the work identified below. The information on this certificate has been made a part of the Copyright Office records.

Maria A. Pallante

Register of Copyrights, United States of America

Registration Number

TXu 1-853-944

Effective date of registration:

February 21, 2013

Title

Title of Work: Experimentation of a Steering Control for a Spaceship

Completion/Publication

Year of Completion: 2008

Author

■ **Author:** Anis Ibrahim Milad

Author Created: text

Work made for hire: No

Citizen of: United States   **Domiciled in:** United States

Year Born: 1948

Copyright claimant

Copyright Claimant: Anis Ibrahim Milad

2956 Sollers Point Road, Baltimore, MD, 21222, United States

Rights and Permissions

Name: Anis Ibrahim Milad

Email: aimilad@msn.com   **Telephone:** 410-282-8708

Address: 2956 Sollers Point Road

Baltimore, MD 21222 United States

Certification

Name: Anis Ibrahim Milad

Date: February 21, 2013

Page 1 of 1

PART 2

YOUTUBE MY NOTES-MILAD-GROUP 1

COPYRIGHT 2017

Different Religion Different Perspective - By A I Milad [[with written explanations]]

We should know that the extremists are always existed in Islam, Christianity, and Judaism. The religions' leaders must have the opportunities to discuss the positive and peaceful verses of the Holy Books in the media (Internet, television, radio, news papers, etc.). There are verses, in all three Holy Books, which must be known as unholy verses and should be brought to the attention of people around the world. The way the Holy Books were written in books were written after many decades Moses, Jesus, and Muhammad were departed to heaven. Mistakes and intention hateful verses were inserted and added to the Holy Books. Please remember the three religions are different perspectives to the same Almighty God. The three Holy Books should be edited to eliminate the unholy and hateful verses. To believe in one religion and to hate and discriminate against the other religions must stop. Harmony and respect to one another among the three religions must be the nature of the next generations around the world.

Angel's Piano - By A I Milad [[with written explanations]]

Angel's Piano is a music composition by Dr. Anis I. Milad. Copyright 2015. The history of Egypt witnesses the brotherhood of the Egyptian Jewish, the Egyptian Christians, and the Egyptian Muslims. In everyday life the Egyptian hospitality manifests itself in everywhere in Egypt. The composer's family consists of Jewish members, Moslem members, Christian members, and Hindu members and there is a strong bondage in his family. A Business Law professor in Egypt said: The religion is the commandment between you and God and the man-made-laws are the rules between you and another man. The Egyptians live in peace and must have a strong military that protect their freedom. Egypt is the heart beat of humanity, historically and forever. Photos by Egypt State Information Service.

Echoes of Abu Simbel Temple - By A I Milad
[[with written explanations]]

Echoes of Abu Simbel Temple is a music composition by Dr. Anis I. Milad. Copyright 2017. Created in the memories of Giovanni Battist Belzoni (November 5, 1778 – December 3, 1823) and Sarah Belzoni (January, 1783 – January 12, 1870) the Archeologists/Egyptologists who discovered Abu Simbel Temple which was built by Ramses II the Egyptian Pharaoh in ancient Egypt. Photo by Egypt State Information System. Information by the US History Channel.

Rescuing the Morning Glory - By A I Milad
[[with written explanations]]

Rescuing The Morning Glory is a music composition by Dr. Anis I. Milad. Copyright 2015. This music composition is created to bring the attention to the unknown history of mankind in Egypt. There are huge pyramids which are buried in the sand throughout Egypt. If the international society which is led by the United States of America and the United Nations join forces to unearth one of these buried pyramids, this pyramid is expected to uncover the extended history of mankind in Egypt beyond the current written history. The composer asks also the private sectors and the governments around the world to contribute to this historic project. The information is based on the research of Dr. Sarah Parcackf at the University of Alabama at Birmingham. Information by USA History Channel.

The Broken Wing - By A I Milad [[with written explanations]]

The Broken Wingl is a music composition by Dr. Anis I. Milad. Copyright 2015. My friend, if you were injured in the war defending your country or even you were injured in soccer game the wheelchair is sometimes a new successful beginning. Explore your mind and become a physicist, mathematician or computer scientist. Doors are open wide for you to enter the extraordinary world that is waiting for you to discover. Paintings by the international artist Merna Amged.

Egyptian Melody - By A I Milad [[with written explanations]]

Egyptian Melody is a music composition by Dr. Anis I. Milad. Copyright 2015. Created in the memories of Tahia Carioca (February 22, 1915 – September 20, 1999) the Egyptian actress and belly dancer. Photo by Al-Ahram Weekly.

Angel's Piano - By A I Milad [[with written explanations]]

Angel's Piano is a music composition by Dr. Anis I. Milad. Copyright 2015. The history of Egypt witnesses the brotherhood of the Egyptian Jewish, the Egyptian Christians, and the Egyptian Muslims. In everyday life the Egyptian hospitality manifests itself in everywhere in Egypt. The composer's family consists of Jewish members, Moslem members, Christian members, and Hindu members and there is a strong bondage in his family. A Business Law professor in Egypt said: The religion is the commandment between you and God and the man-made-laws are the rules between you and another man. The Egyptians live in peace and must have a strong military that protect their freedom. Egypt is the heart beat of humanity, historically and forever. Photos by Egypt State Information Service.

The Americans of Ancient Egypt - By A I Milad [[with written explanations]]

The Americans of Ancient Egypt is a music composition by Dr. Anis I. Milad. Copyright 2015. In the memory of the American Archaeologist/

Egyptologist: James Henry Breasted (August 27, 1865 – December 2, 1935). Photos by Egypt State Information Service.

Absence of Civilization - By A I Milad
[[with written explanations]]

Absence of Civilization is a music composition by Dr. Anis I. Milad. Copyright 2015. Created In the memories of Doria Shafik, Saad Zaghloul, Ester Fanous, and Huda Shaarawi.

Global Warming - By A I Milad [[with written explanations]]

Global Warming is a music composition by Dr. Anis I. Milad. Copyright 2017. Painting by the international artist Merna Amged. It is true that the plants absorb carbon dioxide and release oxygen during the day. The carbon dioxide and the other harmful gases are converted to the trunks and the leaves of the trees which are solid substances. Accordingly, we are not only causing the pollution in our atmosphere but the earth might be absorbing poisoning gases from the space. We can change or adjust the DNA of some of the plants in the most polluted towns to absorb more of the poisoning gases and to convert them to solid substances. (Milad - Copyright 2015)

Neutrality of the Religious Conviction - By A
I Milad [[with written explanations]]

Neutrality of the Religious Conviction is a music composition by Dr. Anis I. Milad. Copyright 2017. Painting by the international artist Merna Amged. There is a historic and global conflict among the three major religions. The Jewish do not accept Jesus as a Messiah and the Christians do not accept Muhammad as a prophet. The conflict is the result of the unholy verses in the three Holy Books which should be edited to reflect the Holy verses only. Although, The Jewish do not accept Jesus as a Messiah and the Christians do not accept Muhammad as a prophet, Jesus said His teachings are the continuation of the Torah and Moses' teachings, and Muhammad in el Koran stated His teachings are the extension of the teachings of Moses and Jesus. And the Old Testament predicted of Jesus the Messiah and the New Testament predicted the coming of prophets who will come after Jesus ((1 Corinthians 14:1-40)). We all were born as a Jewish, Christian, or Moslem and we did not choose to be a Jewish, Christian, or Moslem. We grew up in a certain religion. We must understand that the three religions glorify the same almighty God and every religion is a different perspective of the same almighty God. The unholy and hateful verses in the Three Holy Books must be edited and deleted. Please stop the wars among religions.

YOUTUBE MY
NOTES-MILAD-GROUP 2

COPYRIGHT 2017

Existence of the Living Entity On Earth - By A I Milad [[with written explanations]]

Existence of the Living Entity On Earth is a music composition by Dr. Anis I. Milad. Copyright 2017. Photo by Vickie Milad. We had no choice to be a human being or an ant. The living entity who makes us all alive is an invisible part of our existence. Our body consists of elements, as such iron sodium and etc. The living entity is the creation of the Higher Power of the Holy Spirit that chooses our existence.

To Be Devoted To Eternity - By A I Milad [[with written explanations]]

To Be Devoted To Eternity is a music composition by Dr. Anis I. Milad. Copyright 2017. Photo by Egypt State Information System. Since the dawn of human being's civilization the Man kind worshipped the Almighty God because we are carrying the God gene in our DNA. "A hereditary unit consisting of a sequence of DNA that occupies a specific location on a chromosome and is transcribed into an RNA molecule that may function directly or be translated into an amino acid chain. Genes undergo mutation when their DNA sequences change." https://www.bing.com/search?q=define+gene+and+dna&form=PR USEN&pc=EUPP_&mkt=en-us&httpsmsn=1&refig=5fa43fc07484 42a5afee740c7f2e4857&sp=-1&pq=define+gene+and+dna&sc=1-1- 9&qs=n&sk=&cvid=5fa43fc0748442a5afee740c7f2e4857."

"After comparing more than 2,000 DNA samples, an American molecular geneticist has concluded that a person's capacity to believe in God is linked to brain chemicals." http://www.telegraph.co.uk/news/uknews/1476575/God-gene-discovered-by-scientist-behind-gay-DNA-theory.html

"Dr Hamer insisted, however, that his research was not antithetical to a belief in God. He pointed out: "Religious believers can point to the existence of god genes as one more sign of the creator's ingenuity - a clever way to help humans acknowledge and embrace a divine presence." http://www.telegraph.co.uk/news/uknews/1476575/God-gene-discovered-by-scientist-behind-gay-DNA-theory.html

It is important to know that the Almighty God is known to humanity with different names during the history of humanity; for example, Ra, Jehovah, Yahweh, Father, Son, Holy Spirit, Allah, Shiva, Brahma, Vishnu, and many other names were associated with geographic areas and point out to different characteristics of the Almighty God who created the Universe and Human Beings. The different names also reflect different perspectives to the same Almighty God. Human mind knows that there is God who is the ultimate power in our Universe.

The Holy Spirit My Almighty God - By A I
Milad [[with written explanations]]

The Holy Spirit My Almighty God is a music composition by Dr. Anis I. Milad. Copyright 2017. Photo by Egypt State Information System. The Father (The Almighty Power in the universe), The Son (The Creator and the Destructive Force), and the Holy Spirit (The Comforter)........ They are One Supernatural God. The three characteristics of God is The Almighty Power in the universe, The Creator and the Destructive Force, and The Comforter.

Shiva Nataraja - Lord of the Dance - By A I Milad [[with written explanations]]

Shiva Nataraja - Lord of the Dance is a music composition by Dr. Anis I. Milad. Copyright 2017. Photo by NASA. BBC stated that: Shiva is the third god in the Hindu triumvirate. The triumvirate consists of three gods who are responsible for the creation, upkeep and destruction of the world. The other two gods are Brahma and Vishnu. Brahma is the creator of the universe while Vishnu is the preserver of it. Shiva's role is to destroy the universe in order to re-create it.

http://www.bbc.co.uk/religion/religions/hinduism/deities/shiva.shtml

The characteristics of the three Gods are the same as the characteristics of the Almighty God of the three Holy Books, Old and the new testaments, and El Koran El Kareem. Please remember those Gods are different perspectives to the same Almighty God. Lord Shiva's Ardhanarishvara is reflected in the Old Testament (Genesis 1:27) "So God created human beings making them to be like him. He created them to be male and female." On the other hand, Shiva and our Almighty God are destructive and re-creators. We can remember the destruction of the flood and Noah's Ark to re-create new Human Beings and Sodom's and Gomorrah's destruction. Interesting to know One Almighty God is known by different names.

Saint Karas the Anchorite Pray for Humanity - By A I Milad [[with written explanations]]

Saint Karas the Anchorite Pray for Humanity is a music composition by Dr. Anis I. Milad. Copyright 2017. Painting By the international artist Merna Amged. Saint Karas the Anchorite, also known as Anbba Karas the Tourist (القديس ابنالا ساراك السائح) was a saint of the Coptic Orthodox Church who lived during the late fifth and early sixth centuries. According to his biographer, the Egyptian monk Saint Pambo, he spent 57 years in isolation in the Scetis Desert in communion with God.

207

https://topics.revolvy.com/topic/Saint%20Karas&item_type=topic

St. Karas lived in the late years of the fifth century and the beginning of the sixth century. He was the brother of King Theodosius the great. The great St. Karas led a pure life in the company of Angels, in the Western mountain of scetis. He stayed there 57 years interacting with the Lord. May we learn from St. Karas how to spend that quality time with God away from the world. May his prayers be with us and Glory be to God Forever, amen.

http://www.orthodoxbookstore.org/thebiographyofstkarastheanchorite.aspx

Our Baba Saint Kirollos Pray for Us - By A I Milad [[with written explanations]]

Our Baba Saint Kirollos Pray for Us is a music composition by Dr. Anis I. Milad. Copyright 2017. Painting by the international artist Merna Amged. In the memory of Pope Saint Kirollos VI (Baba Kirollos VI) of Alexandria. He was born Azer Youssef Atta (8 August 1902 – 9 March 1971). Baba Kirollos VI was 116th Pope of Alexandria and Patriarch of the See of St. Mark from 10 May 1959 to his death on 9 March 1971. He was declared a saint after 40 years from his death. This is the time of hurricanes; however, we need to pray to the Holy Spirit Higher Power to be with human being regardless of citizenship, religion, and ethnicity.

Reference: https://topics.revolvy.com/topic/Pope%20Kyrillos%20VI&item_type=topic

YOUTUBE MY
NOTES-MILAD-GROUP 3

COPYRIGHT 2017

The Manners of My Laughing Machine - By A I Milad [[with written explanations]]

The Manners of My Laughing Machine is music by Dr. Anis I. Milad. Copyright 2017. Painting by the international artist Merna Amged. The Manners of My Laughing Machine is the logic in our mind. We are planned creations but we have the choice to believe in our intelligence or not. Our Genes and DNA are telling us there is God DNA which is planned in every human being through the existence of Mankind in every geographic area on earth. For example, when we believe in Yahweh or Allah we are actually believing in the same Almighty God's different characteristics and perspectives not in two different gods. It is important to know that the God's names are taking from the nature previous names. For example, according to James Henry Breasted (August 27, 1865 – December 2, 1935), "the name of Yahweh was the name of the volcano in Sinai-Egypt and was worshiped as a god, also according to Hugo Winckler (4 July 1863 – 19 April 1913) the name of Allah was believed (or not) to be the name of the moon in the Arabian Desert, which is now Saudi Arabia, and was worshiped as a god in the pre-Islamic era." As a result, the teachings of Moses and Mohamed are reflecting the different characteristics and perspectives of the same Almighty God with those names which are associated with those geographic areas. It is important to know that there are huge possibilities that Breasted and Winckler were not accurate in their analyses and their analyses were based on pure speculations.

210

The Holy Spirit My Almighty God - By A I Milad [[with written explanations]]

The Holy Spirit My Almighty God is a music composition by Dr. Anis I. Milad. Copyright 2017. Photo by Egypt State Information System. The Father (The Almighty Power in the universe), The Son (The Creator and the Destructive Force), and the Holy Spirit (The Comforter)........ They are One Supernatural God. The three characteristics of God is The Almighty Power in the universe, The Creator and the Destructive Force, and The Comforter.

Existence of the Living Entity On Earth - By A I Milad [[with written explanations]]

Existence of the Living Entity On Earth is a music composition by Dr. Anis I. Milad. Copyright 2017. Photo by Vickie Milad. We had no choice to be a human being or an ant. The living entity who makes us all alive is an invisible part of our existence. Our body consists of elements, as such iron sodium and etc. The living entity is the creation of the Higher Power of the Holy Spirit that chooses our existence.

To Be Devoted To Eternity - By A I Milad [[with written explanations]]

To Be Devoted To Eternity is a music composition by Dr. Anis I. Milad. Copyright 2017. Photo by Egypt State Information System. Since the dawn of human being's civilization the Man kind worshipped the Almighty God because we are carrying the God gene in our DNA. "A hereditary unit consisting of a sequence of DNA that occupies a specific location on a chromosome and is transcribed into an RNA molecule that may function directly or be translated into an amino acid chain. Genes undergo mutation when their DNA sequences change." https://www.bing.com/search?q=define+gene+and+dna&form=PR USEN&pc=EUPP_&mkt=en-us&httpsmsn=1&refig=5fa43fc07484 42a5afee740c7f2e4857&sp=-1&pq=define+gene+and+dna&sc=1-1-9&qs=n&sk=&cvid=5fa43fc0748442a5afee740c7f2e4857."

"After comparing more than 2,000 DNA samples, an American molecular geneticist has concluded that a person's capacity to believe in God is linked to brain chemicals." http://www.telegraph.co.uk/news/uknews/1476575/God-gene-discovered-by-scientist-behind-gay-DNA-theory.html

"Dr Hamer insisted, however, that his research was not antithetical to a belief in God. He pointed out: "Religious believers can point to the existence of god genes as one more sign of the creator's ingenuity - a clever way to help humans acknowledge and embrace a divine presence." http://www.telegraph.co.uk/news/uknews/1476575/God-gene-discovered-by-scientist-behind-gay-DNA-theory.html

It is important to know that the Almighty God is known to humanity with different names during the history of humanity; for example, Ra, Jehovah, Yahweh, Father, Son, Holy Spirit, Allah, Shiva, Brahma, Vishnu, and many other names were associated with geographic areas and point out to different characteristics of the Almighty God who created the Universe and Human Beings. The different names also reflect different perspectives to the same Almighty God. Human mind knows that there is God who is the ultimate power in our Universe.

The Holy Spirit My Almighty God - By A I Milad [[with written explanations]]

The Holy Spirit My Almighty God is a music composition by Dr. Anis I. Milad. Copyright 2017. Photo by Egypt State Information System. The Father (The Almighty Power in the universe), The Son (The Creator and the Destructive Force), and the Holy Spirit (The Comforter)........ They are One Supernatural God. The three characteristics of God is The Almighty Power in the universe, The Creator and the Destructive Force, and The Comforter.

Shiva Nataraja - Lord of the Dance - By A I Milad [[with written explanations]]

Shiva Nataraja - Lord of the Dance is a music composition by Dr. Anis I. Milad. Copyright 2017. Photo by NASA. BBC stated that: Shiva is the third god in the Hindu triumvirate. The triumvirate consists of three gods who are responsible for the creation, upkeep and destruction of the world. The other two gods are Brahma and Vishnu. Brahma is the creator of the universe while Vishnu is the preserver of it. Shiva's role is to destroy the universe in order to re-create it.

http://www.bbc.co.uk/religion/religions/hinduism/deities/shiva.shtml

The characteristics of the three Gods are the same as the characteristics of the Almighty God of the three Holy Books, Old and the new testaments, and El Koran El Kareem. Please remember those Gods are different perspectives to the same Almighty God. Lord Shiva's Ardhanarishvara is reflected in the Old Testament (Genesis 1:27) "So God created human beings making them to be like him. He created them to be male and female." On the other hand, Shiva and our Almighty God are destructive and re-creators. We can remember the destruction of the flood and Noah's Ark to re-create new Human Beings and Sodom's and Gomorrah's destruction. Interesting to know One Almighty God is known by different names.

Saint Karas the Anchorite Pray for Humanity - By A I Milad [[with written explanations]]

Saint Karas the Anchorite Pray for Humanity is a music composition by Dr. Anis I. Milad. Copyright 2017. Painting By the international artist Merna Amged. Saint Karas the Anchorite, also known as Anbba Karas the Tourist (حئاسلا ساراك ابنالا سيىدقلا) was a saint of the Coptic Orthodox Church who lived during the late fifth and early sixth centuries. According to his biographer, the Egyptian monk Saint Pambo, he spent 57 years in isolation in the Scetis Desert in communion with God.

213

https://topics.revolvy.com/topic/Saint%20Karas&item_type=topic

St. Karas lived in the late years of the fifth century and the beginning of the sixth century. He was the brother of King Theodosius the great. The great St. Karas led a pure life in the company of Angels, in the Western mountain of scetis. He stayed there 57 years interacting with the Lord. May we learn from St. Karas how to spend that quality time with God away from the world. May his prayers be with us and Glory be to God Forever, amen.

http://www.orthodoxbookstore.org/thebiographyofstkarastheanchorite. aspx

Our Baba Saint Kirollos Pray for Us - By A I Milad [[with written explanations]]

Our Baba Saint Kirollos Pray for Us is a music composition by Dr. Anis I. Milad. Copyright 2017. Painting by the international artist Merna Amged. In the memory of Pope Saint Kirollos VI (Baba Kirollos VI) of Alexandria. He was born Azer Youssef Atta (8 August 1902 – 9 March 1971). Baba Kirollos VI was 116th Pope of Alexandria and Patriarch of the See of St. Mark from 10 May 1959 to his death on 9 March 1971. He was declared a saint after 40 years from his death. This is the time of hurricanes; however, we need to pray to the Holy Spirit Higher Power to be with human being regardless of citizenship, religion, and ethnicity.

Reference: https://topics.revolvy.com/topic/Pope%20Kyrillos%20 VI&item_type=topic

Different Religion Different Perspective - By A I Milad [[with written explanations]]

We should know that the extremists are always existed in Islam, Christianity, and Judaism. The religions' leaders must have the opportunities to discuss the positive and peaceful verses of the Holy Books in the media (Internet, television, radio, news papers, etc.).

There are verses, in all three Holy Books, which must be known as unholy verses and should be brought to the attention of people around the world. The way the Holy Books were written in books were written after many decades Moses, Jesus, and Muhammad were departed to heaven. Mistakes and intention hateful verses were inserted and added to the Holy Books. Please remember the three religions are different perspectives to the same Almighty God. The three Holy Books should be edited to eliminate the unholy and hateful verses. To believe in one religion and to hate and discriminate against the other religions must stop. Harmony and respect to one another among the three religions must be the nature of the next generations around the world.

July 16, 2017 @akbatmasr

Angel's Piano - By A I Milad [[with written explanations]]

Angel's Piano is a music composition by Dr. Anis I. Milad. Copyright 2015. The history of Egypt witnesses the brotherhood of the Egyptian Jewish, the Egyptian Christians, and the Egyptian Muslims. In everyday life the Egyptian hospitality manifests itself in everywhere in Egypt. The composer's family consists of Jewish members, Moslem members, Christian members, and Hindu members and there is a strong bondage in his family. A Business Law professor in Egypt said: The religion is the commandment between you and God and the man-made-laws are the rules between you and another man. The Egyptians live in peace and must have a strong military that protect their freedom. Egypt is the heart beat of humanity, historically and forever. Photos by Egypt State Information Service.

Echoes of Abu Simbel Temple - By A I Milad
[[with written explanations]]

Echoes of Abu Simbel Temple is a music composition by Dr. Anis I. Milad. Copyright 2017. Created in the memories of Giovanni Battist Belzoni (November 5, 1778 – December 3, 1823) and Sarah Belzoni (January, 1783 – January 12, 1870) the Archeologists/Egyptologists

215

who discovered Abu Simbel Temple which was built by Ramses II the Egyptian Pharaoh in ancient Egypt. Photo by Egypt State Information System. Information by the US History Channel.

Rescuing the Morning Glory - By A I Milad [[with written explanations]]

Rescuing The Morning Glory is a music composition by Dr. Anis I. Milad. Copyright 2015. This music composition is created to bring the attention to the unknown history of mankind in Egypt. There are huge pyramids which are buried in the sand throughout Egypt. If the international society which is led by the United States of America and the United Nations join forces to unearth one of these buried pyramids, this pyramid is expected to uncover the extended history of mankind in Egypt beyond the current written history. The composer asks also the private sectors and the governments around the world to contribute to this historic project. The information is based on the research of Dr. Sarah Parcackf at the University of Alabama at Birmingham. Information by USA History Channel.

The Broken Wing - By A I Milad [[with written explanations]]

The Broken Wingl is a music composition by Dr. Anis I. Milad. Copyright 2015. My friend, if you were injured in the war defending your country or even you were injured in soccer game the wheelchair is sometimes a new successful beginning. Explore your mind and become a physicist, mathematician or computer scientist. Doors are open wide for you to enter the extraordinary world that is waiting for you to discover. Paintings by the international artist Merna Amged.

Egyptian Melody - By A I Milad [[with written explanations]]

Egyptian Melody is a music composition by Dr. Anis I. Milad. Copyright 2015. Created in the memories of Tahia Carioca (February 22, 1915 – September 20, 1999) the Egyptian actress and belly dancer. Photo by Al-Ahram Weekly.

Angel's Piano - By A I Milad [[with written explanations]]

Angel's Piano is a music composition by Dr. Anis I. Milad. Copyright 2015. The history of Egypt witnesses the brotherhood of the Egyptian Jewish, the Egyptian Christians, and the Egyptian Muslims. In everyday life the Egyptian hospitality manifests itself in everywhere in Egypt. The composer's family consists of Jewish members, Moslem members, Christian members, and Hindu members and there is a strong bondage in his family. A Business Law professor in Egypt said: The religion is the commandment between you and God and the man-made-laws are the rules between you and another man. The Egyptians live in peace and must have a strong military that protect their freedom. Egypt is the heart beat of humanity, historically and forever. Photos by Egypt State Information Service.

The Americans of Ancient Egypt - By A I Milad [[with written explanations]]

The Americans of Ancient Egypt is a music composition by Dr. Anis I. Milad. Copyright 2015. In the memory of the American Archaeologist/ Egyptologist: James Henry Breasted (August 27, 1865 – December 2, 1935). Photos by Egypt State Information Service.

Absence of Civilization - By A I Milad [[with written explanations]]

Absence of Civilization is a music composition by Dr. Anis I. Milad. Copyright 2015. Created In the memories of Doria Shafik, Saad Zaghloul, Ester Fanous, and Huda Shaarawi the Egyptian activists in the beginning of the 20th century.

Global Warming - By A I Milad [[with written explanations]]

Global Warming is a music composition by Dr. Anis I. Milad. Copyright 2017. Painting by the international artist Merna Amged. It is true that the plants absorb carbon dioxide and release oxygen during the day.

217

The carbon dioxide and the other harmful gases are converted to the trunks and the leaves of the trees which are solid substances. Accordingly, we are not only causing the pollution in our atmosphere but the earth might be absorbing poisoning gases from the space climate. We can change or adjust the DNA of some of the plants in the most polluted towns to absorb more of the poisoning gases and to convert them to solid substances. (Milad - Copyright 2015)

Neutrality of the Religious Conviction - By A I Milad [[with written explanations]]

Neutrality of the Religious Conviction is a music composition by Dr. Anis I. Milad. Copyright 2017. Painting by the international artist Merna Amged. There is a historic and global conflict among the three major religions. The Jewish do not accept Jesus as a Messiah and the Christians do not accept Muhammad as a prophet. The conflict is the result of the unholy verses in the three Holy Books which should be edited to reflect the Holy verses only. Although, The Jewish do not accept Jesus as a Messiah and the Christians do not accept Muhammad as a prophet, Jesus said His teachings are the continuation of the Torah and Moses' teachings, and Muhammad in el Koran stated His teachings are the extension of the teachings of Moses and Jesus. And the Old Testament predicted of Jesus the Messiah and the New Testament predicted the coming of prophets who will come after Jesus ((1 Corinthians 14:1-40)). We all were born as a Jewish, Christian, or Moslem and we did not choose to be a Jewish, Christian, or Moslem. We grew up in a certain religion. We must understand that the three religions glorify the same almighty God and every religion is a different perspective of the same almighty God. The unholy and hateful verses in the Three Holy Books must be edited and deleted. Please stop the wars among religions.

YOUTUBE MY
NOTES-MILAD-GROUP 4

COPYRIGHT 2018

Guided By the Holy Spirit - By A I Milad
[[with written explanations]]

Guided By the Holy Spirit is a music composition and an opinion by
Dr. Anis I. Milad. Copyright 2018. Photo by Vickie Milad. "You are not
a human being having a spiritual experience. You are a spiritual being
having a human experience." (Teilhard de Chardin) (French Geologist,
Priest, Philosopher and Mystic, 1881-1955)

Security Systems for Schools the Solution-Updated -
By A I Milad [[with written explanations]]

Security Systems for Schools the Solution-Updated is a music
composition and an opinion by Dr. Anis I. Milad. Copyright 2018.
Painting by the international artist Merna Amged. Security Systems
for Schools the Solution-Updated The Problem: The school
shooting is increasing and the security systems for the school must
be implemented in each school with no exceptions. The Idea: The
schools are now high disaster organizations and a security system
must be implemented to identify the students, to detect metals or to
use revolving door systems. The Solution: The security system such
Swipe Card systems, metal detector systems, and/or revolving door
systems should be applied immediately to all the schools in the United
States. In the revolving door systems could also be for metal detectors.
When the metal is detected the revolving door will stop rotating while
the person who carries the metal is inside the door. In the absent of the
civilization the uncivilized people carry guns to protect themselves but

in the civilized society the civilized people create security systems to prevent the violence and killing. References: Swipe Card Door Locks - buyerzone.com Adwww.buyerzone.com/Access-Control/Swipe-Card 2,700+ followers on Twitter Need a Key Card Swipe Lock System for Your Business? Compare Quotes and Save! buyerzone.com has been visited by 10K+ users in the past month [PDF] Magnetic Swipe Card System Security - University Of Maryland https://www.cs.umd. edu/~jkatz/THESES/... Magnetic Swipe Card System Security A case study of the University of Maryland, College Park Daniel Ramsbrock Stepan Moskovchenko Christopher Conroy Security Metal Detectors Adwww.pti-world.com/metaldetector #1 Walk Through & Hand Held Systems - Lowest Prices on Quality Products Types of Security Metal Detectors. There are two types of security detectors used in. Revolving door systems: https://www.bing.com/images/search?q=.....

Saint Patrick the Christian Missionary - By A I Milad [[with written explanations]]

Saint Patrick the Christian Missionary is a music composition and an opinion by Dr. Anis I. Milad. Copyright 2018. Photo by Wikipedia. "Saint Patrick (Latin: Patricius; Irish: Pádraig [ˈpˠaːɟ ɾˠəʝ]; Welsh: Padrig) was a fifth-century Romano-British Christian missionary and bishop in Ireland. Known as the 'Apostle of Ireland'." https://en.wikipedia.org/wiki/Saint_P...

Reflections on Palm Sunday - By A I Milad [[with written explanations]]

Reflections on Palm Sunday is a music composition and an opinion by Dr. Anis I. Milad. Copyright 2018. Painting by the international artist Merna Amged. "Palm Sunday is a Christian feast day which falls on the Sunday before Easter. It commemorates the triumphal entry of Jesus into Jerusalem in the days before his Passion (Crucifixion), during which his supporters greeted him as the coming Messiah." http://www.newworldencyclopedia.org/e...

No Danger for Humanity - By A I Milad
[[with written explanations]]

No Danger for Humanity is a music composition and an opinion by Dr. Anis I. Milad. Copyright 2018. Photo by NASA. "A study on the Kelly twins, NASA confirms that space travel can change astronauts' genes, even after returning to Earth" (USA TODAY, 2018). As a result of this study the DNA will be changing in case the climate is changing either by the pollution that is caused by the industrial revolution or by the space climate. In both cases human being will survive as consequences of the ability of the DNA to change to adapt to the new climate. Looking for another planet to occupy is less logical than the DNA ability to change and to adapt to a new climate on the same planet.

Replacement of Nuclear Warhead Missiles - By
A I Milad [[with written explanations]]

Replacement of Nuclear Warhead Missiles is a music composition and an opinion by Dr. Anis I. Milad. Copyright 2014 and 2018. Photo by Vickie Milad. This is a wake-up call. It is known that the Department of Defense is working diligently "to achieve a stealthy and survivable subsonic cruise missile........to engage specific enemy warships from beyond the reach of counter-fire systems" (TTO Programs). I would like to add my voice to this urgent technology and also I might say that the American nuclear warhead missiles are ineffective because the anti-missiles are produced by other countries, probably Russia and China, e.g., and the Russian president Putin is much a reminder of Hitler and the Nazi regime when they began to invade the neighboring countries. Converting the American nuclear warhead missiles to stealth nuclear warhead missiles is the most important/urgent technology we need in our time. The United State of America must be sure also the American nuclear warhead missiles not only stealth but at least 12 times faster than the speed of sound. The speed of the Chinese rockets in China is Mach 12.

Airborne Hybrid Vehicle - By A I Milad
[[with written explanations]]

Airborne Hybrid Vehicle is a music composition and an opinion by Dr. Anis I. Milad. Copyright 2013 and 2018. Technical patents by the composer. The Problem: "Satellites today are launched via booster rocket from a limited number of ground facilities, which can involve a month or longer of preparation for a small payload and significant cost for each mission. Launch costs are driven in part today by fixed site infrastructure, integration, checkout and flight rules. Fixed launch sites can be rendered idle by something as innocuous as rain, and they also limit the direction and timing of orbits satellites can achieve." (TTO Programs). The Solution is to combine a rocket and aircraft in one flying vehicle. This improved flying vehicle is equipped to fly in the atmosphere and to orbit the earth in space. The reason this flying vehicle should be hybrid is to be able to orbit the earth and to shorten the time/costs to deliver the Satellite. The Idea: 1) To create an improved flying vehicle (IFV) to carry the satellite to the orbit. 2) The rocket of the IFV will have a new nozzle design. 3) The IFV will be launched from any military base around the globe regardless of the bad weather here or in other part of the world because the IFV will be launched to the earth orbit in a few minutes in a clear sky or not as an ordinary jet aircraft. 4) The IFV will be operated by a pilot(s) who will drive the IFV using jet engines then will ascend to orbit via the rocket. 5) The two jet engines will be turned off and closed from both side while the IFV is in the orbit. 6) The rocket will be fueled for the exact distance from the moment the two jet engines are turned off and the moment the rocket is ignited. 7) The IFV will start by flying upward as an ordinary jet airplane. 8) The rocket will be ignited to overcome the gravity in its way to the orbit. 9) The IFV will be reused indefinitely as long as no maintenance is required. 10) The rocket will be refueled with solid fuel or other suitable fuel. Two options are remained for the IFV which are to release a new satellite or to upload an old satellite for repair or because it is outdated.

Alternative Weapons - By A I Milad
[[with written explanations]]

Alternative Weapons is a music composition and an opinion by Dr. Anis I. Milad. Copyright 2015 and 2018. Photo by NASA. The Problem: 1) Hitting the target and escaping the war zone will not be accomplished easily because there is a strong possibility the enemy's warplanes will follow our warplane and they will shoot it down. 2) Firing ground-to-air missiles against our warplanes will be precise and deadly. 3) Operating unmanned or manned missiles through the satellite or through manual defense systems must need different strategies to lessen the effect of the counter attacks. The Solution: The Strategies are: 1) The first strategy is to equip our warplanes with laser beam which will be aimed toward the enemy's warplane that follows our warplane after hitting the target in the enemy's territory. 2) The purpose of the laser beam is to be used as a destruction force and/ or to blind the pilot of the enemy's warplane. 3) The second strategy is that if firing ground-to-air missiles against our warplanes is done through the satellite and automatically, our warplane must aim high frequency electromagnetic signals against the enemy's defense missiles' location to create electrical interference to the enemy's defense electronic equipment. 4) As another alternative, drones could be specialized to produce high frequency electromagnetic signals by flying in lower altitude than our warplane which is flying in higher altitude to attack the ground enemy location in the war zone. 5) Also the high frequency electromagnetic signals could be used against the enemy's guided missiles to create electrical interference to the enemy's guided missiles electronic system. 6) In the sea and ocean, creating small quiet submarines that are specialized to produce high frequency electromagnetic signals could be the solution to cripple the enemy's quiet submarines when creating electrical interference in the enemy's quiet submarines. 7) Exposing the quiet submarines should be done when launching geographical device to orbit the earth and locate all the submarines in the oceans. This technology is already used for geographical purposes. Taking inventory for all submarines which are in the oceans and the seas is important for protecting this country.

The NAVY will launch the satellite archaeology "Some 400 miles up in space, satellites collect images that are used to identify buried landscapes with astonishing precision." (Bloch, 2013). "University of Alabama at Birmingham archaeologist Sarah Parcak, is a pioneer in using satellite imagery in Egypt." (Bloch). The satellite archaeology is not limited to buried landscapes. 8) Creating drones and small submarines that are equipped/specialized to produce high frequency electromagnetic signals could be the solution to cripple the enemy's quiet submarines, guided missiles, air-to-air missiles and ground-to-air missiles when creating electrical interference in the enemy's electronic offense/defense systems. References Bloch, Hannah (2013) The New Age of Exploration. Satellite Archaeology. Retrieved November 29, 2014 from http://ngm.nationalgeographic.com/201...

Alternatives for Conventional Encounter - By A I Milad [[with written explanations]]

Alternatives for Conventional Encounter is a music composition and an opinion by Dr. Anis I. Milad. Copyright 2013 and 2018. Technical patents by the composer. Description: Helicopter Tank is a combination of a Helicopter and Tank. It consists of helicopter, tank gears, and a removable Howitzer cannon. The Helicopter Tank is designed to fly low, to avoid the radar, and to be driven on the ground in a conventional encounter with the enemy. During the peace time the Helicopter Tank will be used to search for oil and natural gas in the desert and elsewhere. The tank gears are made of light hard aluminum or other strong light weight metal. The transmissions of the horizontal and the vertical propellers will be separated from the transmission of the tank gears. The Idea: During the conventional wars it is necessarily to trick the enemy. The army subunit needs to confuse and to be able to change the position and the location if the fight was during the night. To stay in the same location after a night encounter will expose the army subunit to the enemy air attack in the dawn of the following day especially if the army subunit is a subunit of a howitzers army unit. The loud firing of the howitzers will be another reason for relocating the army subunit after fighting. The Tactic: The army subunit could substitute its all howitzers with only one howitzer which will be attached to the Helicopter Tank which will be relocated in a temporarily location. In this temporarily location the Helicopter Tank will move quite a few yards each time the howitzer fires. The Helicopter Tank that is equipped by the howitzer will be on the move which gives the impression there are several howitzers. When the ammunitions are exhausted the Helicopter Tank will, in the night, be driven on the ground to a safe distance and take off and fly back to the subunit's original location. The soldiers of this subunit will leave fake howitzers which look from a distance as real howitzers expecting the enemy air attack in the dawn. Flexibility: The Captain of the subunit could decide that the howitzer that is attached to the Helicopter Tank be detached and used quietly on the ground to fire the ammunitions or to keep the howitzer attached to the Helicopter Tank while firing it. Shock absorbers will be used for both alternatives. The Helicopter Tank is used instead of a truck to deliver/return the howitzer for the incredible maneuver and speed of the Helicopter Tank.

Egyptians Built the Pyramids - By A I Milad
[[with written explanations]]

Egyptians Built the Pyramids is a music composition and an opinion by Dr. Anis I. Milad. Copyright 2014 and 2018. Photo by Egypt State Information System. "There were no extraterrestrial beings or aliens. The Egyptians definitely built the pyramids and temples. The Lebanese built their million tons columns. The Neolithic, that is, the new stone age was on earth between 10,000-2,000BC and during the ancient Egypt civilization. At this period of time, the earth experienced weak gravity acceleration and force. The Egyptians were able to lift huge and gigantic stones not because of the extraterrestrial machines and equipment but because the gravity acceleration and force were weaker than the gravity acceleration and force that we experience now in this current era." (Milad, 2014). References Milad, A. I. (2014). Neolithic. Copyrighted 2014

Anti-Submarine Warfare - By A I Milad
[[with written explanations]]

Anti-Submarine Warfare is a music composition and an opinion by Dr. Anis I. Milad. Copyright 2015 and 2018. Photo by NASA. The Problem: Delivering the unmanned vessel to the enemy activity zone is slow and not practical. The quiet submarines are not detected in the deep water. The Solution: The solution is to launch geographical device to orbit the earth and locate all the submarines in the oceans. This technology is already used for geographical purposes. Taking inventory for all submarines which are in the oceans and the seas is important for protecting this country. The NAVY will launch the satellite archaeology "Some 400 miles up in space, satellites collect images that are used to identify buried landscapes with astonishing precision." (Bloch, 2013). "University of Alabama at Birmingham archaeologist Sarah Parcak, is a pioneer in using satellite imagery in Egypt." (Bloch). The satellite archaeology is not limited to buried landscapes. The Idea: 1) The geological device in the orbit will allow the Navy to locate all the submarines which are in the oceans and the seas either quiet submarines or not quiet submarines. 2) The location of each submarine continually will be known to the Navy. 3) The quiet submarine will be distinguished and detected because we already know the location of the noisy submarines. 4) Two options are remained. The enemy, quiet or not quiet, submarine will be destroyed by the Air Force or the NAVY warfighters or by the unmanned vessels 5) The stealth unmanned vessels are delivered by the Improved Flying Vehicle or the aircraft carrier if it is approachable. 6) The aircraft carrier will carry aboard the stealth unmanned vessel. 7) The Improved Flying Vehicle (IFV) is equipped to fly in the atmosphere. 8) The IFV will be launched from any military base around the globe regardless of the distance between the military base and the enemy's submarine. 9) The IFV will be operated by a pilot(s) who will drive the IFV away from the war zone or the enemy's submarine faster than the speed of sound after releasing of the stealth unmanned vessels is completed. 10) The unmanned vessel's remote navigator will be directed to follow, detect, or destroy the enemy quiet submarine. 11) The IFV or the aircraft carrier could have one or more unmanned vessels. Reference Bloch, Hannah (2013) The New Age of Exploration. Satellite Archaeology. Retrieved November 29, 2014 from http://ngm.nationalgeographic. com/201...

Anti-Torpedo Submarines - By A I Milad
[[with written explanations]]

Anti-Torpedo Submarines is a music composition and an opinion by Dr. Anis I. Milad. Copyright 2015 and 2018. Photo by NASA. The Problem: A torpedo that is aimed toward an aircraft carrier or another Navy ship must be detected and destroyed before the impact. The Idea: The aircraft carrier, for example, is extremely important during the wars and it is protected by missiles against the enemy warplanes but the aircraft carrier is vulnerable when it comes to the torpedoes. The Solution: 1. Every aircraft carrier must be protected by specialized anti-torpedo submarines. The anti-torpedo submarines will have the capability to dock on both sides of the aircraft carrier. 2. If the anti-torpedo submarines were not docked, the mission is to orbit the aircraft carrier under water to locate incoming enemy torpedoes from a distance away from the aircraft carrier. 3. Both anti-torpedo submarines will continue guarding the aircraft carrier 24/7. 4. The anti-torpedo submarines will be equipped with guided torpedoes which are coded to only destroy the enemy un-coded torpedoes and ships. 5. The anti-torpedo submarines will be carrying stealth torpedoes. 6. The anti-torpedo submarines must be quiet submarines and undetected. 7. The anti-torpedo submarines must be able to sense/detect the enemy torpedoes from all directions including horizontal and vertical incoming enemy torpedoes. 8. During the dock of the anti-torpedo submarines the mission is to replace the used torpedoes with new anti-torpedo torpedoes.

Blackbox the Alternative Solution - By A I Milad [[with written explanations]]

Blackbox the Alternative Solution is a music composition and an opinion by Dr. Anis I. Milad. Copyright 2016 and 2018. Photo by NASA. The Problem: When the airplane crashes in the ocean as a result of an explosion or mechanical reasons, the depth of the ocean and the huge distance of the airplane's pieces from each others will prevent the rescued team for bringing the Blackbox to the surface. Secondly, the battery of the Blackbox is with limited duration. The Idea: "The U.S. Navy operates two extremely low frequency radio (ELF) transmitters to communicate with its deep diving submarines. The Navy's ELF communications system is the only operational communications system that can penetrate seawater to great depths and is virtually jam proof from both natural and man-made interference. It is a critical part of America's national security in that it allows the submarine fleet to remain at depth and speed and maintain its stealth while remaining in communication with the national command authority. The Navy's ELF system operates at about 76 Hz, approximately two orders of magnitude lower than VLF. The result is that ELF waves penetrate seawater to depths of hundreds of feet, permitting communications with submarines while maintaining stealth." (Navy Fact File) Reference: https://fas.org/nuke/guide/usa/c3i/fs... The Solution: 1) Using the extremely low frequency radio (ELF) to communicate with Blackbox. 2) The Blackbox will continue broadcasting the recorded conversation between the pilot and the co-pilot repeatedly when the Blackbox is disconnected from the airplane electrical circles and electrical network.. 3) The Naval Computer and Telecommunications Area Master Station – Atlantic will continue to receive the recorded conversation between the pilot and the co-pilot repeatedly. 4) The Blackbox will be equipped with battery and a broadcasting equipment. 5) This solution will eliminate the search for the Blackbox in the ocean.

Landmine Removal - By A I Milad [[with written explanations]]

Landmine Removal is a music composition and an opinion by Dr. Anis I. Milad. Copyright 2015 and 2018. Photo by Egypt State Information System. The Problem: "Difficult terrain and threats such as ambushes and Improvised Explosive Devices (IEDs) can make ground-based transportation to and from the front line a dangerous challenge. Helicopters can easily bypass those problems but present logistical challenges of their own, and can subject flight crew to different types of threats. They are also expensive to operate, and the supply of available helicopters cannot always meet the demand for their services, which cover diverse operational needs including resupply, fire-team insertion and extraction, and casualty evacuation." Unpractical Solution: "The flare is safe to handle and easy to use. People working to deactivate the mines – usually members of a military or humanitarian organization – simply place the flare next to the uncovered land mine and ignite it from a safe distance using a battery-triggered electric match. The flare burns a hole in the land mine's case and ignites its explosive contents. The explosive burns away, disabling the mine and rendering it harmless." http://www.nasa.gov/home/hqnews/1999/.... It is an unpractical solution because the landmines are located in unknown locations. The Solution of this research: The components of this research are: 1) It is entirely possible to create static electricity, and even lightning using this method. Van de Graaf generators, for example, use rubbing to generate voltages in excess of a 1,000,000V. However, it's a very inefficient method for generating power. Dynamo generators (the standard generator) are surprisingly efficient. http://www.askamathematician.com/2010... 2) Fuse is a component of the landmine. Fuse is a flammable material which is defined as following: "Fuse - A combustible material used to ignite an explosive charge. http://science.howstuffworks.com/land.... The Strategy: 1) Creating a high voltage artificial lightning using dynamo. 2) Installing the dynamo in a helicopter which will be used for this purpose only. 3) While the artificial lightning is produced, the helicopter will be flying in a safe altitude over the landmine field. There is no need to locate the landmines. 4) The artificial lighting will trigger/ignite the combustible material which will ignite the detonator which will ignite the larger/ remaining amounts of explosive. 5) The artificial lightning will continue to be produced and the explosions of the landmines will continue. The helicopter could repeat theses procedures to be sure all the landmines are exploded in the intended zone. 6) During the war, the helicopter could create a quick safe road which can make the ground-based transportation to and from the front line a safe operation.

The Scientists Under the Microscope - By A I Milad [[with written explanations]]

The Scientists Under the Microscope is a music composition and an opinion by Dr. Anis I. Milad. Copyright 2015 and 2018. Photo by NASA. The unscientific science is in the media 24/7 and the scientists are repeating the incorrect information constantly. Here are some examples: The sun's surface is developing a black hole. There are millions of the earth-like planets. Mars is our future planet. We are looking for a planet which can sustain life. From my observations here are the true outlooks: The reality is that the black spot on the surface of the sun is the formation of heavy/solid elements matters which will be discharged in an orbit around the sun and will form a new planet. There is no earth-like planet in the universe because the earth is the unique creation of our sun and it was discharged from our sun. The human is the evolution of one-cell which is produced in the core of the earth and found in the volcanoes' lava that is more than 2000 Fahrenheit. It is a well known scientific fact that the universe is expanding. At any moment, hours or a thousand year, the earth will be moved farther from the sun and approximately to Mars orbit and the domino effect in the Solar System will take its course from Mercury to Neptune. The Earth will be frozen in the new distance and Venus will be cooling and it will be the possible planet for life. Mars will be moving farther in the Solar System and its environment will not be suitable for life. It is a matter of time. We need to study Venus and to monitor this planet because it is our, as a human species, only hope to survive in this universe. We will need to change our DNA to fit the other planet's environment. Looking for a planet that is identical to the earth is not possible. The earth is created by our sun. When the sun nuclear interactions produce heavy/solid elements the sun will discharge these elements in the orbit around the sun. These heavy/solid elements and other gases will generate/construct a new planet. The earth is the sun's creation.

USA War Tactics - By A I Milad [[with written explanations]]

USA War Tactics is a music composition and an opinion by Dr. Anis I. Milad. Copyright 2014, 2015 and 2018. Photo by NASA. I think the war tactics in the 21st century should be changed. For example, if the president of China is giving a speech to declare war against the USA in the Tiananmen Square, the new USA war tactics should bomb the Tiananmen Square and the Chinese president before the Chinese president finishes his sentence. Here is how: NASA should be a future defense organization. It is a mistake to discontinue the production of the space shuttles. Instead of discontinuing the production of the space shuttles, the space shuttle should be improved and armed with atomic warheads. The distance from the ground to the orbit around the earth will approximately take 2 minutes. The shuttle in the orbit will reach China, for example, in approximately 3 minutes. If the Chinese president declares war against the USA from Tiananmen Square we will need to respond swiftly and instantly and the armed space shuttle will accomplish this goal before the Chinese president finishes his sentence.

Laylat al-Qadr - By A I Milad [[with written explanations]]

Laylat al-Qadr is a music composition and an opinion by Dr. Anis I. Milad. Copyright 2018. Photo by Egypt State Information System. Laylat al-Qadr is a music composition which is composed in the memory of my Aunt Mary who stated that she saw the heaven open up in the sky and the angels were hovering in the opening space. It was known that Laylat al-Qadr is only an Islamic Holiday that indicates an Islamic event. Laylat al-Qadr is the day which Mohammed the Prophet received the first verses in the Koran el Kareem. As a Christian, I asked

Aunt Mary this is an Islamic event and you are a Christian woman so how do you explain your story? She looked at me and she said that she does not know!!!! I believe that Judaism, Christianity, and Islam, beside other religions around the world, are each describing different characteristics of the same Almighty God. At this time, Aunt Mary and I did not know that the heaven opened up twelve times in the past and the events were recorded in the new and old testaments. Reference: https://bible.knowing-jesus.com/topic...

Seeing Heaven Open Up - By A I Milad
[[with written explanations]]

Seeing Heaven Open Up is a music composition by Dr. Anis I. Milad. Copyright 2018. Painting by the international artist Merna Amged. In the memory of Aunt Mary who witnessed this event. Aunt Mary experienced the heaven open up which is also mentioned 12 times in the New and Old Testaments. The scientists believe that the New an Old Testaments and other religions are not scientific books. In my opinion, the New and Old Testaments and other religions are scientific books and must be considered scientific books because they are above human understanding. If individuals witness the heaven open up in a certain moment it means the nature which we are seeing everyday is not a complete picture of the nature.

Defense Alternatives For Navy Ships - By A
I Milad [[with written explanations]]

Defense Alternatives For Navy Ships is a music composition and an opinion by Dr. Anis I. Milad. Copyright 2016. Photo by the NAVY Official Website: http://www.navy.mil/undersec/index.asp Defense Alternatives For Navy Ships The Problem: For sinking Navy Ships the struggle for the ship crew members is deadly. USS Indianapolis is the example of the devastating 4 days which the crew members were eating by the sharks. The Idea: The first alternative is to combine the submarine advantage with the destroyers or any other type of the Navy ships. The second alternative is to detect the enemy torpedoes by a high frequency radar. The First Solution: 1) The Navy Ships must be able to resist the sudden sinking as a result of the enemy torpedo attacks. 2) Combining submarine/destroyer structures must be done to prevent the damaged ship from sinking. 3) Individual/separated air cylinders that will be able to float the wreckage of the damaged ship must be installed permanently in the inner structures of all the Navy ships. 4) The air cylinders must be in protective structures throughout the inner cavity of the Navy ship. 5) The crew of the destroyed Navy ship will be able to hold in the wreckage, the worst scenario, until the help arrives. The Second Solution: 1) By detecting the enemy torpedoes by a radar which is operated by a high frequency radio the Navy ships will be able to fire anti-torpedoes guided missiles ship-to-torpedoes against the enemy torpedoes. 2) Because the radar's waves are high frequency waves then transmitting the high frequency in the distance of water to detect the enemy torpedoes will not be possible. 3) Locating/placing the radar equipment to detect the enemy torpedoes on the four top sides of the Navy ship, that is, on the top front, the top rear, the top right side, and the top left side of the navy ship will be the solution to overcome the inability of transmitting the high frequency in the distance of water. 4) The radar will be locating the enemy torpedoes and the speed. 5) The enemy torpedoes travel a few inches from the ocean surface. 6) As usual the high frequency radar will be used from the top of the Navy ship aimed toward the possible enemy torpedo attacks to detect them. 7) Anti-torpedo guided missiles ship-to-enemy torpedoes will be equipped to locate the enemy torpedoes according to the high frequency radar's instructions.

Adjustable Rocket Engine - By A I Milad
[[with written explanations]]

Adjustable Rocket Engine is a music composition and an opinion by Dr. Anis I. Milad. Copyright 2015. Adjustable Rocket Engine The Problem: There are four concepts must be addressed: 1. The first concept is that the speed/maneuver of the rocket engine is not controllable from the start when the rocket engine is ignited and when the rocket engine replaces the jet engine in a warplane. 2. The second concept is associated with the fighter jet, that is, the limited speed of the fighter jet and the huge distance between the American military bases and the location of the enemy. 3. The third concept is that the response of the fighter jet to the satellite signals (pictures and location) is slow when the satellite locates a particular location of the military equipment/convoy of the enemy because the slow response is due to the huge distance between the USA military bases and the location of the enemy. 4. The fourth concept is that the fighter jet in general confirms limited capability of the pilot as a human being and the maximum/limited speed of the fighter jet. The Solution: 1. The main concern is to narrow the response time between the signals of the satellite and destroying the enemy. The speed of Blackbird is Mach 3 but the speed of the rocket is Mach 10 (China is experimenting with Mach 12). 2. When producing the rocket in the same way the Blackbird is produced to be able to carry several bombs/guided missiles/clusters, intelligent data, stealth, and satellite networking connections the delivery of the bombs for several enemy locations will be faster and instant. 3. The suggested warplane rocket will be unmanned so the physical pilot factor will be eliminated. 4. As an alternative, the warplane rocket will be operated remotely via the satellite. 5. As another alternative, the warplane rocket will be operated by a pilot who will be wearing a space suit and inside a rocket design cockpit. 6. The rocket engine which is suggested by this proposal for this warplane rocket must be equipped with adjustable pumps to control the speed of the rocket and its maneuvers (please see the drawing).